Organic Nonlinear Optical
Materials and Devices

MATERIALS RESEARCH SOCIETY
SYMPOSIUM PROCEEDINGS VOLUME 561

Organic Nonlinear Optical Materials and Devices

Symposium held April 6–9, 1999, San Francisco, California, U.S.A.

EDITORS:

B. Kippelen
The University of Arizona
Tucson, Arizona, U.S.A.

H.S. Lackritz
Gemfire Corp.
Palo Alto, California, U.S.A.

R.O. Claus
Virginia Tech
Blacksburg, Virginia, U.S.A.

Materials Research Society
Warrendale, Pennsylvania

This work was supported in part by the Office of Naval Research under Grant Number ONR-N00014-99-1-0368. The United States Government has a royalty-free license throughout the world in all copyrightable material contained herein.

Single article reprints from this publication are available through University Microfilms Inc., 300 North Zeeb Road, Ann Arbor, Michigan 48106

CODEN: MRSPDH

Published by:

Materials Research Society
506 Keystone Drive
Warrendale, PA 15086
Telephone (724) 779-3003
Fax (724) 779-8313
Web site: http://www.mrs.org/

Library of Congress Cataloging-in-Publication Data

Organic nonlinear optical materials and devices : symposium held April 6–9, 1999,
 San Francisco, California, U.S.A. / editors, B. Kippelen, H.S. Lackritz, R.O. Claus
 p.cm.—(Materials Research Society symposium proceedings,
 ISSN 0272-9172 ; v. 561)
 Includes bibliographical references and index.
 ISBN 1-55899-468-8
 1. Optoelectronic devices—Materials—Congresses. 2. Photonics—Materials—
Congresses. 3. Organic compounds—Congresses. 4. Polymers—Congresses.
5. Nonlinear optics—Materials—Congresses. I. Kippelen, B. II. Lackritz, H.S.
III. Claus, R.O. IV. Series: Materials Research Society symposium proceedings ;
 v. 561
TA1750.0733 1999   99-37090
621.381'045—dc21   CIP

Manufactured in the United States of America

CONTENTS

*Invited Paper

PART III: ELECTRONIC AND LIGHT-EMITTING MATERIALS

*Invited Paper

DEDICATION

In memory of Bruce A. Reinhardt

Bruce A. Reinhardt
1949-1999

Shortly before the 1999 MRS Spring Meeting, Bruce A. Reinhardt passed away. Bruce was an important figure in the community of organic nonlinear optical materials. With his sudden loss, the community lost a great scientist, a colleague, and a friend. To honor his memory, we dedicate the symposium and these proceedings to Bruce.

Bruce Alan Reinhardt was born on March 19, 1949 in Dayton, Ohio. He received his M.S. degree in organic chemistry from Wright State University in 1974. In 1977, Bruce joined the University of Dayton Research Institute and began working on the synthesis and preliminary characterization of curable enyne-containing thermoplastics. Cured composites of these materials were shown to have increased resistance to commercial organic paint stripping solvents when compared to state-of-the-art thermoplastics.

Bruce accepted a position in the Polymer Branch of the Nonmetallic Materials Division as a research chemist in 1978. His initial research involved the synthesis and characterization of reactive diluents for fabrication of difficult-to-process acetylene-terminated phenylquinoxaline thermoset resins. During this time period, Bruce also carried out research on novel polyimides, resulting in unique structures with outstanding thermal stability, and the modification of toughened acetylene-terminated resins, as well as new solvent-resistant polyphenylquinoxaline thermoplastics.

In 1985, Bruce was promoted to synthetic group leader and at that time began directing research on soluble polyphenylene rigid-rod molecules. Bruce's research efforts over the past decade centered on the synthesis of novel organic nonlinear optical chromophores. He directed systematic studies of structure/property relationships to gain a fundamental understanding of the contribution of the electronic structure of the molecules to their NLO behavior. Bruce published over 60 journal articles, and was an author or co-author on over 20 patents. He received the Materials Directorate Cleary Award for scientific excellence in 1991. Because of his technical achievements and amiable, but industrious personality, Bruce Reinhardt will be greatly missed on both a personal and professional basis.

PREFACE

This volume consists of the proceedings of Symposium F, "Organic Nonlinear Optical Materials and Devices," held April 6–9 at the 1999 MRS Spring Meeting in San Francisco, California.

The purpose of this symposium was to provide an interdisciplinary forum for the discussion of recent advances in research on electronic and photonic devices made with organic and polymeric materials. The field of organic optical materials is rapidly growing, and new advances are being made both in attaining a deeper understanding of device phenomena and in designing improved materials for thin films, fibers, and waveguides. The symposium provided a highly interactive forum for an update in the connected areas of photonics with organic materials. The Meeting was highlighted by several major advances in various fields ranging from nonlinear absorbers and electro-optic polymers, to photorefractive polymers, organic transistors, and electroluminescent materials and devices for displays. The symposium highlighted developments in materials chemistry and physics relevant to such devices and struck a balance between basic science and technology.

B. Kippelen
H.S. Lackritz
R.O. Claus

June 1999

ACKNOWLEDGMENTS

The editors would like to thank the Office of Naval Research, AlliedSignal, Lightwave Microsystems, Lockheed Martin, and the Materials Research Society for their generous support. We also gratefully acknowledge the staff at MRS for their efficient and friendly efforts at making the symposium and these proceedings a good experience.

MATERIALS RESEARCH SOCIETY SYMPOSIUM PROCEEDINGS

MATERIALS RESEARCH SOCIETY SYMPOSIUM PROCEEDINGS

Prior Materials Research Society Symposium Proceedings available by contacting Materials Research Society

Part I

Nonlinear Optical Materials

MULTIFUNCTIONAL FULLERENE FILMS FOR PHOTONICS

Y. LIU *, R. O. CLAUS **, Y.X. WANG **, H. LU **, T. DISTLER *
*NanoSonic, Inc., 200 Country Club Drive, A-2, Blacksburg, VA 24060, nano@usit.net
**Fiber & Electro-Optics Research Center, Virginia Tech, Blacksburg, VA 24061-0356

ABSTRACT

We report here the preparation of highly homogeneous multifunctional fullerene thin films for photonics using the electrostatic self-assembled monolayer technique at room temperature for the first time. The monolayers and multilayers were characterized via contact angle measurements, UV-vis spectroscopy and atomic force microscopy. These results demonstrated that close-packed, highly uniform fullerene or fullerene/metal cluster films with micron-thickness could be formed on various substrates including silicon, glass, metal and plastics.

INTRODUCTION

Buckminstrerfullerene (C_{60}) is a hollow cage, all carbon molecule with 60 carbon atoms connected by a conjugated backbone. This molecule is comprised of twelve pentagons and twenty hexagons [1]. Fullerenes including C_{60} and C_{70} have attracted tremendous interest among chemists, physicists and material scientists due to their unique physical and chemical properties since their discovery. Large second order [2], and third order nonlinear optical responses [3], photoconductivity [4], and photorefractivity [5] of the fullernenes and their charge-transfer complexes [6] have been reported. Extensive studies have been carried out focusing on the preparation of uniform thin film and sturctural characterization of these molecules. The Langmuir-Blodgett technique has been actively pursued over the last decade, unfortunately, film prepared by this method still suffer problems such as visible non-uniform patches and inhomogeneities [7]. A few papers have been published concerning use of the self-assembly monolayer (SAM) technique to construct a few layers of thin fullerene films [8]. Thermally evaporated fullerene films are normally deposited at elevated temperatures and in high vacuum environments, however, the film uniformity and thickness are very hard to control and not repruducible [9].

In this paper, we will report an alternate strategy to produce high quality multilayer fullerene and fullerene/metal cluster films – the use of an electrostatic self-assembly multilayer (ESAM) technique, for the first time [10]. It will be demonstrated that micron-thick and molecular-level homogeneous fullerene and fulerene/Pt cluster films can be fabricated under standard environment conditions. There are several distinguishing features to our approach, 1) the assembly method is extremely simple, 2) the thickness, index of refraction, and functionality (electronic, optical, etc.) of multimonolayer segments may be precisely controlled, by incorporation of selected molecules and by process control, 3) the room temperature and pressure solution process allows the formation of fullerene films on substrates of arbitrary sizes and geometries.

EXPERIMENT

The fullerenes, a C_{60} and C_{70} mixture (fullerite, 9:1), were purchased from Aldrich and used as a precursor for water-soluble fullerenes. The water-soluble fullerene, fullerol tetramethylammonium salt, was synthesized in our lab.

3

Platinum nanoparticles protected by PDDA surface passivation coatings were prepared by the hydrogen reduction method. Potassium tetrachloro platinate (K_2PtCl_4, Aldrich) dissolved in an aqueous solution containing PDDA with a molar ratio of Pt : polymer of 1 : 5 was reduced by bubbling first with argon, and then hydrogen gases, vigorously, each for 15 minutes. A change of color from light yellow to dark brown was immediately observed. The typical particle size about 1-2 nm, is revealed by transmission electron microscopy (EM-420, Philips) after the solution is stored for 4 months.

Single crystal silicon (p-111) wafers were obtained from ITME (Poland) and cleaned based on the RCA cleaning procedures. First, the substrates were immersed in a mixture of H_2O/H_2O_2 (30%)/NH_4OH (29 w/w% as NH_3) (5: 1: 1) for 30 minutes and washed with water. Afterwards the substrates were dipped into a mixture solution of H_2O_2/H_2SO_4 (96% w/w) (3: 7) for 10 minutes, followed by rinsing with water. The aqueous poly (diallyldimethylammonium chloride) (PDDA, high molecular weight, 20% w/w in water, Aldrich) solution was diluted to 0.5% v/v. The concentration of the aqueous fullerol solution used is 1.0×10^{-5} mol/L (pH 7.8).

The typical procedure of the deposition films has been described earlier [11]. In brief, a pre-cleaned slide (negatively charged at suitable pH range, i.e. pH 7 for silica or silicon) was immersed in cationic PDDA solution for a minute, then rinsed extensively with water. Subsequently, the slide was dipped into the anionic fullerol solution for a minute, followed by thorough washing with water. The cycles can be repeated as many times as required for the construction of multilayer films. The deposition of Pt/fullerene films was carried out in a similar way.

Contact angles were measured on a NRL C.A. Goniometer 100-00 115 (Rame-Hart, Inc.) using the sessile drop method. The volume of a drop was 4 μL. The sessile drops of nanopure water were carefully placed on the film surfaces with a microliter syringe held in an upright position. Contact angles were determined within 1 min of applying the drops to the film. All reported values are the average of at least six measurements taken on both sides of at least three drops. UV-vis spectroscopy (Hitachi U-2010) was used to monitor the film deposition process. The regular increase in the absorption spectrum is due to the layer-by-layer increment in the amount of material deposited on the substrate. Noncontact mode atomic force microscopy images were collected employing a Metris-2000NC AFM (Burley, Instruments, Inc.).

RESULTS AND DISCUSSIONS

The spontaneous, alternating layer-by-layer self-assembly of the cationic PDDA and anionic fullerene molecules is based on the electrostatic attraction developed between the oppositely charged species, which promotes strong interlayer adhesion and a uniform multilayer deposition process. Due to the strong hydrophobic force among the fullerene molecules, dilute solutions with concentrations less than 10^{-5} g/mol is recommended. The water contact angle measurements monitoring the self-assembly process are shown in Figure 1.

The contact angle for the PDDA monolayer on single crystal silicon is 24^0. When the anionic fullerol adsorbs on this surface, the contact angle is increased to 46^0. The increased angle is attributed to the adsorption of fullerol with polar head groups attached to the positively charged surface and the nonpolar hydrocarbon or carbon chains oriented upward toward the bulk portion of the water droplet. As the process continues, i.e. another layer of PDDA adsorbs on the fullerol surface, the contact angle decreases to 33^0 due to hydrophilic property of the cationic polyelectrolyte, but slightly higher than that of PDDA directly on silicon surface, which indicates the sublayer molecules influence the hydrophobic properties of the outermost layer at some degree. As more layers of fullerol and PDDA molecules are built up, the average contact angle oscillates between 35^0 and 47^0 according to which molecules form the outermost layer, which clearly demonstrates that alternating layers of anionic fullerol and cationic fullerol molecules can be fabricated in a layer-by-layer fashion.

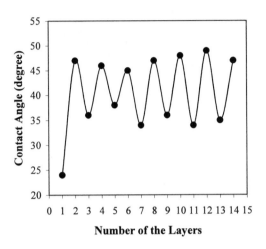

Figure 1. Contact angles of PDDA/Fullerol ESAM films as function of the number of layers.

The UV-vis absorption spectra of PDDA/fullerol ESAM films are shown in Figure 2. A linear deposition pattern of a 80-bilayer film is observed with respect to the number of bilayers, yielding an optical density of 0.0323 per bilayer at 256 nm. The linear nature of the plot suggests that each layer adsorbed contributes an equal amount of material to the thin film. Several PDDA/fullerol ESAM films with as many as 200 bilayers have been fabricated on quartz, silicon and glass slides although there is no practical limit to the deposition of more layers.

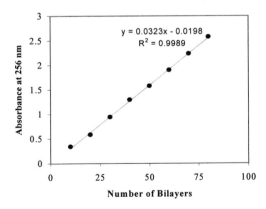

Figure 2. Optical absorbance of PDDA/Fullerol ESAM films via the number of bilayers at 256 nm.

: 73.71 A

X: 900.00 nm   Y: 900.00 nm

Figure 3. Atomic force microscopy image of one layer of fullerol on one layer of PDDA.

Direct observation of the atomic arrangement and surface morphology/topography of the grown fullerol films have been studied using atomic force microscopy. The AFM images of one layer of a fullerol film on a one-layer PDDA modified single crystal silicon substrate is shown in Figure 3. The observed island-like structures having average diameter of about 20 nm are typical for such films as the fullerol concentration used is 4×10^{-5} g/mol. We have found that the concentration of fullerol solution plays some role in determining the diameters of the fullerol clusters. The more dilute the concentration of fullerol, the less aggregate the fullerol clusters, and the smaller the diameter of the fullerol clusters, which also results in thinner and more uniform films.

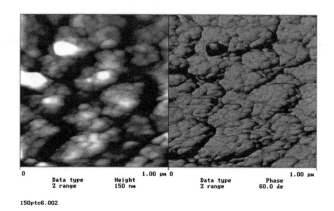

0   1.00 µм 0   1.00 µм
   Data type   Height   Data type   Phase
   Z range   150 nm   Z range   60.0 de

150ptc6.002

Figure 4. Tapping-mode AFM image of a 150 bilayers of ESAM Pt:PDDA/fullerenol film deposited on a single crystal silicon substrate.

Atomic force microscopy (AFM) analysis of the Pt:fullerene films was performed under ambient conditions using a Nanoscope IIIa SPM (Digital Instruments) operated in the tapping-mode. The topography and phase AFM micrographs of a 150 bilayer Pt:fullerene film are shown in Figure 4. Examination of the images suggest several features of the Pt:fullerene film: (i) the Pt nanoparticles and fullerene clusters are closely packed, (ii) a buckyball shaped structure with an average diameter of about 20 nm is observed for the films, and (iii) the average surface roughness is about 32 nm. It is interesting to note that there is little or no visible organic-inorganic phase separation.

CONCLUSIONS

Monolayer and multilayer thin films of fullerene and fullerene/Pt clusters have been fabricated using the ESAM method for the first time. This technique is based on the strong electrostatic force existed among the adjacent opposite-charging molecules, and the film is built up in a layer-by-layer fashion with nanometer-level uniformity. By judicious selection of functional molecules and metal or inorganic clusters, our study has provided us the opportunity to build micron-thick and multifunctional fullerene film which can find wide applications in optical limiting, second and third nonlinear optical devices, charge-transfer complex (*i.e.*, fullerene/Pt, and fullerene/PPV), photoconductor (for example, fullerene/PVK) and photorefractive devices. Additional results of our studies of these novel fullerene films are forthcoming.

ACKNOWLEDGMENTS

This work was supported by Air Force contract F33615-98-C-5421, Office of Naval Research contract N00014-96-1-1285, and National Science Foundation grants DMI-9860084.

REFERENCES

1. H. W. Kroto, J. R. Heath, S. C. Obrien, R. F. Curl, and R. E. Smalley, Nature **318**, 162 (1985); R. C. Haddon, L. E. Brus, and K. Raghavachari, Chem. Phys. Lett. **125**, 459 (1986).
2. X. K. Wang, T. G. Zhang, W. P. Lin, S. Z. Liu, G. K. Wong, M. M. Kappes, R. P. H. Chang, and J. B. Ketterson, Appl. Phys. Lett. **60**, 810 (1992).
3. H. Kanbara, T. Maruno, A. Yamashita, S. Matsumoto, and T. Hayashi, J. Appl. Phys. **80**, 3674 (1996).
4. Y. Wang, Nature **356**, 585 (1992).
5. Y. Zhang, Y. Cui, and P. N. Prasad, Phys. Rev. B **46**, 9900 (1992).
6. P. V. Kamat, J. Am. Chem. Soc. **113**, 9705 (1991); J. R. Sension, A. Z. Szarka, G. R. Smith, R. M. Hochstrasser, Chem. Phys. Lett. **185**, 179 (1991).
7. Y. S. Obeng, A. J. Bard, J. Am. Chem. Soc. **113**, 6279 (1991); J. Guo, Y. Xu, Y. Li, C. Yang, Y. Yao, D. Zhu, C. Bai, Chem. Phys. Lett. **195**, 625 (1992); Y. Tomioka, M. Ishibashi, H. Kajiyama, and Y. Taniguchi, Langmuir, **9**, 32 (1993); M. Maggini, A. Karlsson, L. Pasimeni, G. Scorrano, M. Prato, L. Valli, Tetrahedron, Lett. **35**, 2985 (1994); M. Matsumoto, H. Tachibana, R. Azumi, M. Tanake, and T. Nakamura, Langmuir, **11**, 660 (1995).
8. K. Chen, W. B. Caldwell, C. A. Mirkin, J. Am. Chem. Soc. **115**, 1193 (1993); J. A. Chupa, S. Xu, r. F. Fischetti, R. M. Strongin, J. P. jr. McCauley, A. B. Smity, J. K. Blasie, J. Am. Chem. Soc. **115**, 4383 (1993); J. Z. Zhang, M. J. Geselbracht, A. B. Ellis, J. Am. Chem. Soc. **115**, 7789 (1993).
9. A. F. Hebard, Annu. Rev. Mater. Sci. **23**, 159 (1993).
10. A) R. Iler J. Colloid Interface Sci. **21**, 569 (1966); b) G. Decher, and J. D. Hong Ber. Bunsen-Gen. Phys. Chem. **95**, 1430 (1991); c) W. S. Keller, H. N. Kim, and T. E. Mallouk, J. Am. Chem. Soc. **116**, 8817 (1994); d) M. Ferreira, M. F. Rubner, Macromolecules **28**, 7101 (1995); e) Y. Liu, A. Wang, and R. O. Claus, Appl. Phys. Lett. **71**, 2265 (1997).
11. Y. Liu, R. O. Claus, J. Appl. Phys. **1**, 1 (1999).

LANGMUIR-BLODGETT FILMS OF SQUARYLIUM DYE J-AGGREGATES EXHIBITING FEMTOSECOND OPTICAL RESPONSES

M. FURUKI*, H. KAWASHIMA**, T. TANI***, L. S. PU***
* The FESTA Laboratories, 5-5 Tokodai, Tsukuba, Ibaraki 300-26, Japan, furuki@festa.or.jp
** Electrotechnical Laboratory, 1-1-4 Umezono, Tsukuba, Ibaraki 305, Japan
***Corporate Res. Labs., Fuji Xerox Co. Ltd., 430 Sakai, Nakaimachi, Kanagawa 259-01, Japan

ABSTRACT

J-aggregates of squarylium dyes in Langmuir films have been found to exhibit highly efficient and ultrafast nonlinear optical properties. We established a novel method for making Langmuir-Blodgett (LB) films of J-aggregates with a single absorption band at 784 nm. Deposition of the LB-films was carried out under excess compression of the Langmuir films at a constant speed. Coating the surface of the LB-film with a glassy poly-perfluorocarbon was found to enhance and stabilize the formation of J-aggregates. Characterization of morphology using a near-field scanning optical microscope (NSOM) showed that this LB-film has a highly ordered structure comprising 2-dimensional domains of J-aggregates. We also observed nonlinear-optical responses from this LB-film. Ultrafast decay of absorption change (250 fs) with quite low saturation energy (3.4 $\mu J/cm^2 \cdot$pulse) was observed in femtosecond pump-probe measurements. These results suggest highly delocalization of excited states in the J-aggregates of this LB-film with the 2-dimensional mono-molecular layer structure which initially formed in the Langmuir film.

INTRODUCTION

Increasing attention has been paid to the nonlinear optical properties of excitons, not only in inorganic materials, but also in organic materials. The J-aggregate of dye molecules is a particularly interesting topic in this area [1-8]. Enhancements of the optical nonlinearity and fast decay of excited states in J-aggregates have been discussed in the picture of delocalized Frenkel excitons [2-8]. Previous studies concerned the J-aggregates of uncertainly dispersed forms in some matrixes at low temperature. Since the NSOM was demonstrated [9], the structures of aggregation forms have been able to discuss in terms of their optical properties at spatial resolutions down to several tens nm [10]. Practical applications strongly require the observation of highly efficient and ultrafast nonlinear optical responses at room temperature from a film of J-aggregates with a well-defined structure.

In our previous study on squarylium dye Langmuir film, molecular structures of the dye forming J-aggregates in the Langmuir film were specified [11]. The Langmuir film showed an intense J-band at about 780 nm and was found to exhibit highly efficient and ultrafast nonlinear optical properties even in the form of a mono-molecular layer at the air-water interface [12-14]. In the present study, a novel method for depositing LB-films of the squarylium dye J-aggregates was established. Structural characterization using the NSOM and observation of femtosecond nonlinear optical responses by resonant pump-probe measurements are discussed in this paper.

EXPERIMENT

In fabricating LB-films, the stability of aggregation forms against compression on the water surface is important. We chose SQ33mMe as shown in Fig. 1 for the material of LB-films. A moving-wall type LB-film deposition apparatus (NL-LB240-MWA, Nippon Laser & Electronics Lab.) was used to prepare the LB-Film. A solution of squarylium dye was

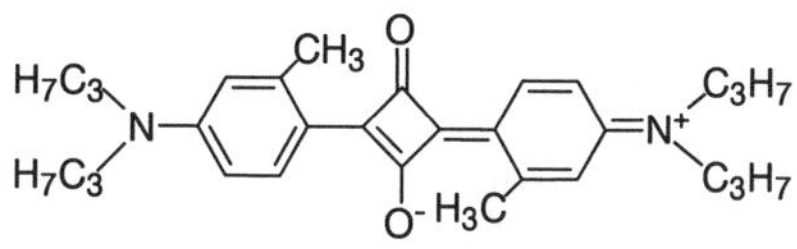

Fig. 1. Chemical formula of SQ33mMe

prepared by dissolving SQ33mMe in chloroform at a concentration of 8.31×10^{-4} M (i.e., 20 $\mu l = 1 \times 10^{-16}$ molecules), and was spread on the surface of pure water containing 0.1 wt% of NaCl with the temperature controlled to be 5 °C. Glass slides having a width of 50 mm the same as the LB-trough were used as the substrates. To stabilize the deposited film, it was coated with glassy fluorocarbon polymer (Cytop, Asahi Glass). To investigate the morphology of the J-aggregate, transmission images at a wavelength of 780 nm were obtained using a optical microscope with a band-pass filter and NSOM for Langmuir films and LB-films, respectively. In the NSOM imaging, a Topometrix Aurora was employed in illumination mode with a laser diode coupled to the fiber-tips. In order to stabilize scanning of the fiber-tip and to reduce optical noise, we modified some instruments used in share-force feedback. The femtosecond excited-state dynamics were investigated by measuring the transient absorption change spectra using pump pulses with a center wavelength of 784 nm and probe pulses of a white light continuum with cross-correlation width of 200 fs under a 1 kHz repetition rate. The diameter of both the pump and probe beams was 5 mm².

RESULTS AND DISCUSSION

After a chloroform solution of SQ33mMe was spread on the water surface, a Langmuir film of J-aggregates was formed. Figure 2 shows the π-A isotherm and corresponding absorption spectra of the Langmuir film. Initially, a single J-band at 784 nm was observed. This band was split into two bands, one of which was slightly blue-shifted to 770 nm, by compression of the Langmuir film in accordance with the increase in surface pressure.

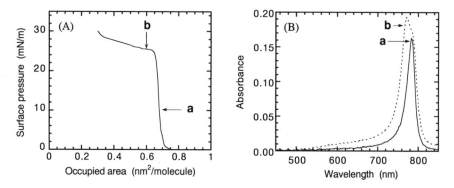

Fig. 2. (A) Π-A isotherm and (B) corresponding absorption spectra of a SQ33mMe Langmuir film.

Deposition of LB-films normally requires compression of the Langmuir film and is done under condition of constant surface-pressure. In the case of this Langmuir film, deposition under a constant surface-pressure of 20 mN/m results in decomposition of J-aggregates in the process of film transfer as shown in Fig. 3 (a). On the other hand, deposition of LB-films maintaining the form of J-aggregates can be achieved by excess compression of the Langmuir-film beyond monolayer occupation. Withdrawing substrate begin when the molecular occupation area reached to 0.6 nm² at a withdrawing speed of 5 mm/min. under a constant compression speed of 7 mm/min., resulting in deposition of a LB-film exhibiting a similar absorption spectrum (Fig. 3 (b)) to the Langmuir film (Fig. 2(b)). Because SQ33mMe has only short alkyl-substituents, the Langmuir film has lower elasticity than other materials conventionally used for LB-films. Therefore, the requirement of excess compression may be attributed to an irrbalance in the surface-pressure of the Langmuir film throughout the compressed area. In order to stabilize this J-aggregate, we coated the surface of the LB-film with hydrophobic polymer film (Cytop). Not only stabilization, but also realignment of molecules was observed as shown in Fig. 3 (c). After the surface coating, the deposited LB-film was sandwiched between a hydrophilic glass-substrate and a hydrophobic polymer film. Stretching of the deposited monolayer caused by an increase in interfacial tension can explain the acquisition of a LB-film with a single J-band which similar to the spectrum observed on a water-surface under low surface-pressure (Fig. 2 (a)).

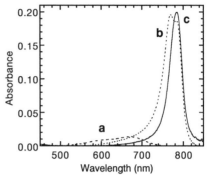

Fig. 3. Absorption spectra of deposited LB-films (a) under constant surface pressure, (b) under constant compression and (c) with coating.

LB-films with a single J-band are produced by deposition with a constant compression and post surface coating. It is important to characterize the morphology of such LB-films in comparison with that of the Langmuir film. In transmission micrographs of the Langmuir film obtained using polarized light at 780 nm (resonant to the J-band) as shown in Fig. 4, we can observed that the film comprises 2-dimensional domains of J-aggregates. The darkness of each domain depends on the difference in angle between the polarization of incident light and the orientation of transition dipole-moment in J-aggregates. Because each domain has one axis of transition dipole-moment at the J-band, images with opposite contrast are observed when polarized light of two orthogonal directions is used.

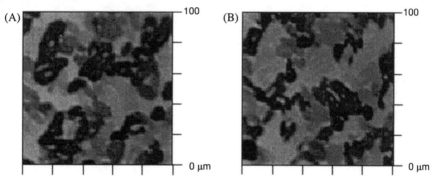

Fig. 4. Transmission micrographs of 2-dimensional J-aggregates in a SQ33mMe Langmuir -film under illumination by light of 780 nm and detection with a polarizer in the (A) vertical and (B) horizontal directions.

In the case of the deposited LB-film, characterization of the morphology was performed by NSOM. Figure 5 shows the topography (A) and two transmission images (B, C) of the LB-film. The topography suggests that the LB-film is flat at a resolution on the order of nm. The thickness of the coated polymer film is found to be about 20 nm by measuring the height of steps formed by holes in the coating observed in the same image. In the two transmission images obtained using polarized illumination of two orthogonal directions, opposite contrasts were observed. Although the size of each single domain was reduced into sub μm levels compared to that of Langmuir film observed in Fig. 4, these images suggest that the molecular arrangement is uniform within each domain of J-aggregates in the LB-films.

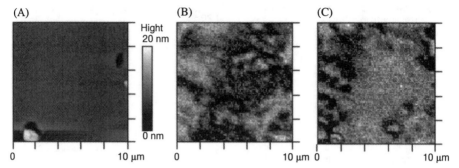

(A)   (B)   (C)

Hight
20 nm

0 nm

0   10 μm   0   10 μm 0   10 μm

Fig. 5. (A) Topography and transmission images of a SQ33mMe LB-film observed by NSOM
with illumination by polarized laser light of 780 nm in (B) vertical and (C) horizontal directions.

The linear optical properties of absorption spectra and morphology of domains observed in
transmission images by NSOM show that the LB-films preserve 2-dimensional structure of J-
aggregates. Characterization of nonlinear optical properties was conducted by resonant pump-
probe measurement using femtosecond laser pulses. Figure 6 shows the results of pump-probe
measurements of different durations with a pump energy of 3.0 μJ / cm^2 · pulse. As in the case of
conventional of the J-aggregates [2-8], dispersion-type transient spectra with a decrease in the
absorbance at longer wavelengths and an increase at shorter wavelengths appeared upon
illumination by the pump pulses. Reductions in these absorbance changes were observed within
several hundred femtoseconds, suggesting a fast response of the J-aggregates in the LB-films. To
analyze the decay lifetime of excited states, temporal change in absorption bleaching is plotted in
Fig. 7 using the value of 790 nm, which was the valley wavelength in the absorption change
spectra. Using curve-fitting with two exponential decay time-constants, a fast component of 250
fs (66%) and a slow component of 11 ps (34%) were obtained. As these time constants for the
decay of excited states are similar to those in the Langmuir films [12-14], the fast component may
be attributed to coherent effects of stimulated emission or re-orientation of excited states and the
slow component to the collective radiative-decay process called superradiance.

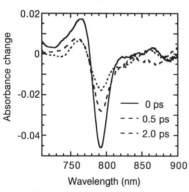

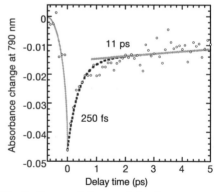

Fig. 6. Transient absorption-change spectra
of the SQ33mMe LB-film with a pump
energy of 3.0 μJ/cm^2•pulse.

Fig. 7. Temporal change of absorbance at
790 nm on the delay time with fitted curves
of two exponential decay time-constants.

Figure 8 shows the dependence of maximum absorbance change on pump energy. Saturation of bleaching was observed with a pump energy of over 5 μJ/cm²•pulse. The saturation value of bleaching was about 0.06, which is the one-third of the initial absorbance at 790 nm. According to the Frenkel-type exciton model of J-aggregates [2-8], dispersion-type transient-absorption spectra must be the sum of a negative-absorption change caused by bleaching of linear absorption and a positive-absorption change by induced transition from the 1-exciton states to the higher exciton states at slightly shorter wavelength. In the case of this LB-film, J-band has a large band width of about 35 nm (FWHM), because of inhomogeneous and homogeneous broadening in solid films at

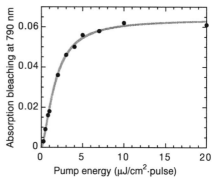

Fig. 8. Dependence of absorption bleaching on pump energy.

room temperature. Therefore, both the negative and the positive values will overlap each other in a wide spectral range. This explains why the bleaching does not reach the whole value. From the curve in Fig. 8, the pump energy for absorption saturation is calculated to be 3.4 μJ/cm²•pulse, which is twice the value of half-saturation at 1.7 μJ/cm²•pulse.

By comparing the spectral widths of pump and absorption spectra, the excited fraction was estimated to be 15/35. The saturation energy of 3.4 μJ/cm²•pulse was therefore re-evaluated as 7.9 μJ/cm²•pulse. The ratio of absorbance calculated from the overlap of integrated pump and absorption spectra is 0.34. The number of photons in one pump pulse concerning on excitation of the J-aggregates is 1.05×10^{13} cm^{-2}, which corresponds to an absorption cross-section of 9.5 nm²/photon. On the other hand, the area occupied by one molecule in LB-films calculated from the occupied area based on the surface-pressure vs. area isotherm of Fig. 2 (A) is 0.70 nm²/molecule. Hence the number of 14 molecules determined from division of the absorption cross-section by the occupied area suggest the average of coherently cooperating number of dye molecules in this J-aggregate. Although, the degree of homogeneity of the absorption band and the domain size observed by NSOM are less than those of the Langmuir films. Those results of femtosecond pump-probe measurements clearly indicate the delocalization of excited states. The difference between the coherent area of excitation (9.5 nm²) and the single domain size of the same orientation (sub-μm diameter), can be explained as follows. For the J-aggregates, size of ordered orientation means single domain observed by NSOM, and only a part of the single domain is excited over the coherent area by one photon as the results of femtosecond pump-probe measurements. If the size of the ordered orientation is less than that of the coherent area, the number of cooperation molecules is limited in the single-domain and the absorption band should be blue-shifted.

CONCLUSIONS

LB-films of squarylium dye with a single J-band were produced by deposition with constant compression and post surface coating by a poly-perfluorocarbon film. Inhibition of environmental water adsorption and increase in interfacial tension caused by coating were effective in preserving the 2-dimensional structures of J-aggregates. These effects were also confirmed in the observations by NSOM. LB-films were covered with domains of monolayer J-aggregates in which transition dipole-moments were oriented in the same direction. In femtosecond pump-probe measurement, the LB-films exhibited fast decay of absorption saturation and low saturation energy almost equal to those of the Langmuir films [12-14]. Such a high performance in terms of nonlinear optical properties, comparable to that of a monolayer at the air-water interface, is indicative of highly ordered structure in these LB-films. Such a nonlinear optical response with only a monolayer of dye molecules at room temperature is due to the highly ordered 2-dimensional structure. These results are expected to open the way to construction of planar all-optical switching devices.

ACKNOWLEDGMENTS

This work was supported by the New Energy and Industrial Technology Development Organization (NEDO) in the framework of the Femtosecond Technology Project. We would like to thank Dr. Shynsuke Kobayashi, Dr. Fumio Sasaki, Dr. Tsuyoshi Kato of the Electrotechnical Laboratory, and Dr. Osamu Wada and other members of the FESTA Laboratories for stimulating and fruitful discussions.

REFERENCES

1. E. E. Jelly, Nature **138**, 1009 (1936).
2. H. Fidder, J. Knoester, and D. A. Wiersma, Chem. Phys. Letters **171**, 529 (1990).
3. J. R. Durrant, J. Knoester, and D. A. Wiersma, Chem. Phys. Letters **222**, 450 (1994).
4. F. C. Spano and S. Mukamel, J. Chem. Phys. **95**, 7526 (1991).
5. B. Kopainsky and W. Kaiser, Chem. Phys. Lett. **88**, 357 (1982).
6. S. Kobayashi and F. Sasaki, Nonlinear Optics **4**, 305 (1993).
7. K. Minoshima, M. Taiji, K. Misawa, and T. Kobayashi, Chem. Phys. Lett. **218**, 67 (1994).
8. K. Minoshima, M. Taiji, A. Ueki, K. Miyano, and T. Kobayashi, Nonlinear Optics **14**, 39 (1995).
9. E. Betiz, J. K. Trautman, T. D. Herris, J. S. Weiner, and L. R. Kostelak, Science **251**, 1468 (1991).
10. D. A. Higgins and P. F. Barbara, J. Phys. Chem. 99, 3 (1995).
11. M. Furuki, S. Kim, L. S. Pu, H. Nakahara, and K. Fukuda, J. Chem. Soc. Japan Chem. Ind. Chem. **10**, 1121 (1990).
12. M. Furuki, L. S. Pu, F. Sasaki, S. Kobayashi, and T. Tani, Mat. Res. Soc. Symp. Proc. **488**, 777 (1998).
13. M. Furuki, L. S. Pu, F. Sasaki, S. Kobayashi, and T. Tani, Appl. Phys. Lett. **72**, 21, 2648 (1998).
14. M. Furuki, L. S. Pu, F. Sasaki, S. Kobayashi, and T. Tani, Mol. Cryst. Liq. Cryst. **316**. 67 (1998).

ENHANCEMENT OF NONLINEAR OPTICAL PROPERTIES THROUGH SUPRAMOLECULAR CHIRALITY

S. VAN ELSHOCHT, B. BUSSON, T. VERBIEST, M. KAURANEN, J. SNAUWAERT, L. HELLEMANS, A. PERSOONS

Laboratory of Chemical and Biological Dynamics, University of Leuven, Celestijnenlaan 200D, B-3001 Heverlee, Belgium.

C. NUCKOLLS, T.J. KATZ

Department of Chemistry, Columbia University, New York, NY 10027, USA.

ABSTRACT

We study second-order nonlinear optical properties of enantiomerically pure and racemic Langmuir-Blodgett films of a chiral helicenebisquinone. Supramolecular aggregation of the material enhances the nonlinear efficiency of the enantiomerically pure material. We also show that the material is potentially useful for constructing quasi-phase-matched structures for frequency conversion.

INTRODUCTION

Even order nonlinear optical processes require that the material be noncentrosymmetric [1]. Noncentrosymmetry on the microscopic scale is relatively easy to achieve, for example, through proper design of systems where donor and acceptor groups are connected by a Π-conjugated bridge. Noncentrosymmetry on the macroscopic scale can be achieved for example by the Langmuir-Blodgett (LB) technique [2-3], electric field poling [4] or crystal growth [5].

Another possibility to achieve noncentrosymmetry is the use of chiral materials. Chiral molecules have no mirror plane and exist in two forms that are mirror images of each other (enantiomers). Most physical properties of the two enantiomers are similar, however they differ in optical properties. Chiral molecules are inherently noncentrosymmetric due to the lack of a mirror plane [6]. This results in a second-order nonlinear optical response even in highly symmetric systems such as isotropic solutions.

The use of chiral molecules can be indirect through crystallization into a noncentrosymmetric group [7]. On the other hand, in this paper we show an example where chirality is used directly through supramolecular stacking of enantiomeric molecules into helicene fibres with an enhanced nonlinearity as a consequence.

We studied the tetrasubstituted helicenebisquinone shown in Fig. 1 [8-9]. Former studies have shown that bulk samples of the enantiomerically pure (nonracemic) material consist of long fibres that can be seen with an optical microscope [10]. The racemate (50/50 mixture of the enantiomers) however does not organize into such fibers. We found that Langmuir-Blodgett (LB) films of the racemic and nonracemic materials also have different nonlinear optical (NLO) properties. The second harmonic intensity of a LB-film of the nonracemic material is a factor of 1000 higher than that of a comparable film of the racemic

15

mixture. This factor corresponds to a factor of 30 in the second order nonlinear susceptibility. In addition, we show that the second harmonic signal is dominated by tensor components that are only allowed by chirality.

Our study also suggests that this material could be used to make quasi-phase-matched structures [11]. In such structures, efficient frequency conversion is achieved by periodically alternating the sign of the nonlinearity.

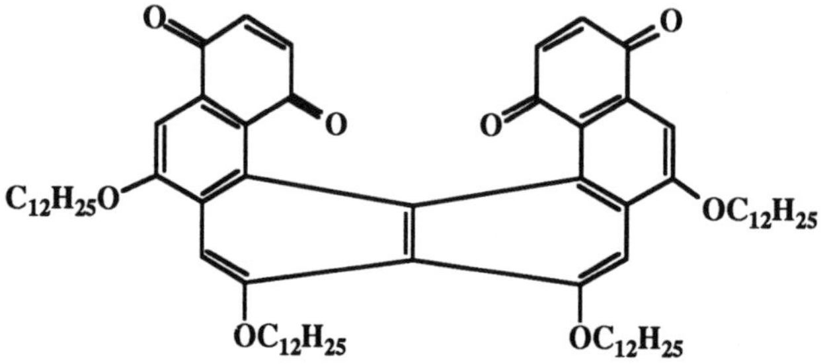

Fig.1 Chemical structure of the tetrasubstituted helicenebisquinone.

EXPERIMENT

We made Langmuir-Blodgett films of the racemic and nonracemic materials. The materials were spread out of a chloroform solution onto the water subphase. The monolayers were compressed to a surface pressure of 20 mN/m and after stabilization transferred onto a glass substrate. The substrates were treated with octadecyltrichlorosilane to make them hydrophobic.

We studied the nonlinear optical properties by irradiating the samples at 45° incidence with a Nd:YAG-laser (1064 nm, 50 Hz, 8ns) and detecting the generated light at the double frequency (532 nm).

The quality of the LB-films was checked by measuring the second harmonic intensity as a function of the number of layers. The signal increased quadratically with the number of layers, indicating good film quality. We made LB-films with the horizontal (up to 10 layers) and vertical (up to 300 layers) dipping technique. At no point was deterioration of the film quality observed.

The intensity of the second-harmonic signal generated by the nonracemic LB-film was a factor 1000 higher than that generated by a racemic LB-film with the same number of layers. This result is rather unexpected because all molecules in both films have the same chemical structure. In addition, for the nonracemic sample the p-in-s-out (fundamental beam p-polarized, SH signal s-polarized) signal was the largest, whereas for the racemic sample the p-in-p-out signal was the largest. The p-in-s-out signal is associated with tensor components that are only allowed in the presence of chirality. On the other hand, the p-in-p-out signal is allowed for all surfaces, chiral and achiral.

Therefore, a possible explanation for the higher nonlinear efficiency of the enantiomeric sample is that the second harmonic signal is dominated by the chiral tensor components. In the racemic sample these tensor components must cancel because the chiral components of the two enantiomers have the same magnitude but opposite sign. Another explanation could be that the helicene molecules pack differently in the enantiomeric pure and racemic sample.

To test these two hypotheses, we measured the s- and p-polarized components of the second harmonic generated (SHG) light as a function of the polarization of the fundamental laser light. Analysis of these polarization patterns allows us to determine the different tensor components of the second order nonlinear susceptibility [12].

The second harmonic intensity can be written as [13]:

$$I(2\omega) = \left| f \cdot E_p^2 + g \cdot E_s^2 + h \cdot E_p \cdot E_s \right|^2 \qquad (1)$$

where E_p and E_S are the s- and p-polarized components of the fundamental field, and f, g, and h are the expansion coefficients which depend on the components of the second-order susceptibility tensor. To check the in-plane isotropy of the sample, the SHG signal is measured while rotating the sample around the surface normal [14].

The racemic samples were isotropic. However, for the nonracemic samples the SH intensity as a function of the angle of rotation of the sample gave a pattern which shows that the samples have C_2 symmetry. In additon, for no angle of rotation did the SH signal vanish. This is an additional indication of the presence of chirality, because for an achiral sample the signal has to go to zero for some angle.

The C_2 symmetry complicates the analysis and interpretation of the experimental results because none of the tensor components can be associated uniquely with chirality. To circumvent this difficulty we had to consider combinations of tensor components that are independent of the rotation about the surface normal. For example, the combination $\chi_{xyz} - \chi_{yxz}$ is rotationally invariant and can be associated with chirality. The magnitude of this combination was determined by comparison with a quartz wedge ($d_{11} = 0.3$ pm/V) [15] and was estimated to be 50 pm/V, which is quite respectable for a material that is not optimized for NLO. Analysis of our measurements shows that the chiral combinations are a factor of ten larger than the achiral ones. To be certain that chirality is the sole factor responsible for the large NLO response, we compared the p-in-p-out signal for a racemic and nonracemic sample. The signal for the nonracemic sample was averaged over the angle of rotation. Both signals were of the same order of magnitude which shows that the achiral contributions of both samples were the same.

To determine the origin of the large chiral contribution to the NLO response, we studied the samples with linear absorption spectroscopy, X-ray diffraction and atomic force microscopy (AFM). The results from the different techniques indicate that the molecules in both the racemic and nonracemic samples aggregate on a microscopic level. The absorption spectra of both samples are similar to those of solutions in which the molecules aggregate. However AFM pictures show that for the nonracemic samples the columnar stacks of molecules organize further into bundles, whereas for the racemic samples the organization remains on a smaller scale. In the LB-films of the racemate, the R and S enantiomers are likely to be segregated [16], but the incompatibility of the two enantiomers probably prevents larger structures from forming. This result is consistent with X-ray diffraction data. No diffraction was observed from the racemic samples, probably because they are less organized.

Thus, the large chiral contributions seem to originate from the organization of the molecules into large helical stacks with large chiral tensor components. Although these stacks are seemingly also present in the racemic films, their chiral tensor components cancel, for that of one enantiomer is the opposite of that of the other.

A possible application of the chemical and optical properties of the helicenebisquinone lies in the design of quasi-phase-matched devices for nonlinear frequency conversion. When SHG processes take place in bulk materials (i.e. much thicker than the films used so far in this study), the fundamental and second-harmonic beams progressively lose their phase relation as they propagate because of dispersion effects. After one coherence length, they are out of phase. The efficiency of conversion is then very low and does not increase with the thickness of the material. The phase relation can be restored, and continuous growth of the SHG output intensity achieved, by reversing the sign of the susceptibility coefficients $\chi^{(2)}$ after one coherence length (lowest-order quasi phase matching) or an odd multiple (higher-order quasi phase matching). This is traditionally done by alternating dipole alignment of a polar material. We show here that chirality provides a new scheme for achieving quasi phase matching without polar ordering.

In chiral materials, the tensor components associated with chirality change sign between the two enantiomers, whereas the achiral components maintain their sign. Structures consisting of alternating layers of the two enantiomers can therefore provide a new approach to sign inversion requirements for quasi phase matching.

To test the compatibility of the enantiomers in alternating structures, we prepared a series of multilayer LB-films comprised of both enantiomers. The films were made of units of four layers of a single enantiomer. We had two different units as building blocks: one unit composed of the R enantiomer and one of the S enantiomer. As such, films were made with up to eigth units, i.e., 32 layers. We used either the same units (R/R/R/...) or alternated the units (R/S/R/...). Second-harmonic light was detected in transmission where the coherence length is much larger than the thickness of the samples. The nonlinear response of the units can therefore be added together. Because of the chirality of the helicene, for the first type of films the signal should increase quadratically with the number of layers. On the other hand, the signal should vanish if the number of R units is equal to the number of S units [Fig. 2].

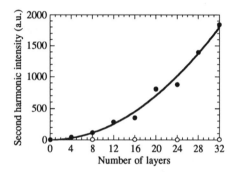

Fig. 2 Second harmonic intensity of the LB-films as a function of the number of layers: R/S/... structures (open dots) and R/R/... structures (filled dots). The solid line is a quadratic fit of the data points for the R/R/... structures.

The first evidence of a quasi-phase matching (QPM) effect was obtained by studying a set of samples prepared by layering different numbers of layers of one enantiomer on an equal number of layers of the other. If the experiments had been conducted by detecting transmitted light, the number of layers required would have been very large. To avoid having to deposit so many layers, we instead detected the second harmonic light that was reflected, for the coherence length is then very short [17]. The SHG intensity as a function of the thickness was calculated assuming the QPM effect, taking into account the absorption of the second harmonic beam by the sample. The absorption coefficient was that measured directly by UV-visible spectroscopy. The experimental data perfectly match the theoretical curve (Fig. 3). In particular, the output intensity is observed to decrease when the stacks are thicker than the coherence length. The value for the coherence length deduced from the best fit corresponds to 48.5 molecular layers per stack. This work is being refined with larger numbers of stacks.

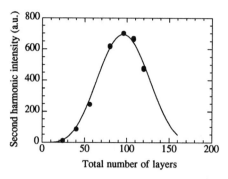

Fig. 3 Phase matching curve for the R/S structure. The solid line represent a theoretical fit to the experimental datapoints.

CONCLUSIONS

In this paper we have shown that the intensity of the second harmonic light, generated by an LB film of the enantiomer of the helicenebisquinone, is a factor of 1000 more intense compared to a similar film made from the racemic mixture. In addition, we have demonstrated that the explanation is found in the chiral supramolecular stacking of the enantiomer molecules. Finally, we have demonstrated that chirality offers a new scheme to design quasi-phase-matched structures.

ACKNOWLEDGEMENTS

We acknowledge financial support from the Belgian government, the Belgian National Science Foundation, and the Katholieke Universiteit Leuven. T.V. is a postdoctoral fellow and L.H. is a research associate of the Fund for Scientific Research-Flanders. M.K. acknowledges the support of the Academy of Finland. C.N. and T.J.K. thank the Kanagawa Academy of Science and Technology and NSF for support.

REFERENCES

[1]   P.N. Prasad and D.J. Williams, *Introduction to Nonlinear Optical Effects in Molecules and Polymers*, Wiley, New York, 1991.
[2]   G. Roberts, *Langmuir-Blodgett films,* Plenum, New York, 1990.
[3]   A. Ulman, *An introduction to ultrathin organic films from Langmuir-Blodgett to self-assembly,* Academic Press, San Diego, 1991.
[4]   S. Miyato and H. Sasabe, *Advances in nonlinear optics vol.4 : Poled polymers and their applications to SHG and EO devices,* Gordon and Breach Publishers, Singapore, 1997.
[5]   Ch. Bossard, K. Sutter, Ph. Prêtre, J. Hulliger, M. Flörsheimer, P. Kaatz, and P. Günter, *Advances in nonlinear optics vol.1 : Organic nonlinear optical materials,* Gordon and Breach Publishers, Singapore, 1995.
[6]   P.M. Rentzepis, J.A. Giordmaine, K.W. Wecht, Phys. Rev. Lett. **16**, 792 (1966).

[7]   J. Zyss and D.S. Chemla, *Nonlinear optical properties of organic molecules and crystals,* D.S. Chemla ans J. Zyss Eds. Vol.1, Academic Press, Orlando, 1987.
[8]   C. Nuckolls, T.J. Katz, L. Castellanos, J. Am. Chem. Soc. **118**, 3767 (1996).
[9]   A.J. Lovinger, C. Nuckolls, T.J. Katz, ibid. **120**, 264 (1998).
[10]   C. Nuckolls, T.J. Katz, G. Katz, P.J. Collins, L. Castellanos, in preparation.
[11]   M.M. Fejer, G.A. Magel, D.H. Jundt, R.L. Byer, IEEE J. Quantum Electron. **28**, 2631 (1992).
[12]   M. Kauranen, J.J. Maki, T. Verbiest, S. Van Elshocht, A. Persoons, Phys. Rev. B **55**, R1985 (1997).
[13]   J.J. Maki, M. Kauranen, and A. Persoons, Phys. Rev. B **51**, 1425 (1995).
[14]   X. Zhuang, D. Wilk, L. Marrucci, Y.R. Shen, Phys, Rev. Lett. **75**, 2144 (1995).
[15]   D. Roberts, IEEE J. Quantum Electron. **28**, 2057 (1992).
[16]   P. Nassoy, M. Goldmann, O. Boulossa, F. Rondelez, Phys. Rev. Lett. **75**, 457 (1995).
[17]   Ref.: J. U. Kang, Y. J. Ding, W. K. Burns and J. S. Melinger, Opt. Lett. **22**, pp 862-864 (1997).

BRIDGED POLYSILSESQUIOXANES WITH IMPROVED SECOND ORDER NONLINEAR OPTICAL PROPERTIES AND STABILITY

S. T. HOBSON[a], J. ZIEBA[b], P. N. PRASAD[b], AND K. J. SHEA[c]
[a]Drug Assessment Division, U. S. Army Med. Res. Inst. of Chem. Def., APG, MD, 21010-5405*
[b]Department of Chemistry SUNY Buffalo, Buffalo, NY 14260-3000
[c]Department of Chemistry University of California, Irvine, California, 92697-2025

ABSTRACT

We report the synthesis and sol-gel polymerization of 4-nitro-N,N-bis[(3-triethoxysilyl)propyl]aniline **1**. An efficient synthesis of the monomer was developed by the hydrosilylation of N, N-diallyl-4-nitroaniline. Optical quality thin films were synthesized by spin coating an *n*-butanol solution of **1** using formic acid as catalyst and source of water. We improved the temporal stability of the NLO signal from films prepared from 4-nitro-N,N-bis[(3-triethoxysilyl)propyl]aniline by increasing the intensity of the poling field and extending the heating period during the poling/curing stage. By Maker fringe analysis, a $\chi^{(2)}$ value of 9 x 10^{-8} esu was measured for these polysilsesquioxanes. If one assumes that the major component of the NLO effect is along the z-axis, the $\chi^{(2)}$ value corresponds to a d_{33} coefficient of 18.9 pm/V and a r_{33} value of 4.7 pm/V.

INTRODUCTION

In the design of nonlinear optical (NLO) active materials, systems that consist of organic chromophores hold an advantage over inorganic materials because the NLO effect is from electronic transitions of each chromophore, rather than from the bulk response of the material.[1] Although this fact allows the response time of organic chromophores to be on the order of picoseconds, the processing of organic crystals into devices has proven to be problematic.[2] In contrast, the ease of processing polymeric systems containing NLO chromophores into components holds a great potential advantage over organic or inorganic crystals. However, polymeric systems can require high temperatures during fabrication and may suffer from matrix instability. Since the sol-gel process allows the synthesis of rigid network hybrid organic/inorganic materials at room temperature, it is logical therefore to extend sol-gel chemistry into the field of NLO active materials.[3] These efforts have met with some success.

Our own effort approached the incorporation of chromophore into a silicate matrix using a two-point connection in analogy to organic polymers (i.e., main chain linear polyimides[4] and polyurethanes[5]). This concept has been applied to the synthesis of bis(triethoxysilyl) monomers **1-3** using the *para*-nitroaniline (PNA) chromophore (Figure 1).[6]

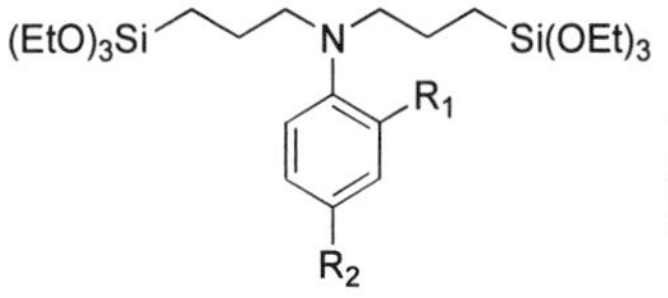

	R_1	R_2
1	H	NO_2
2	NO_2	NO_2
3	CH_3	NO_2

Figure 1. Bis(triethoxysilyl) monomers used for NLO films.

Mat. Res. Soc. Symp. Proc. Vol. 561 © 1999 Materials Research Society

We prepared optical films by sol gel hydrolysis and polycondensation of the α,ω bis(triethoxysilyl) bridged monomer 1 with formic acid as catalyst and water source. After aging for 10 to 30 minutes, the homogeneous solution was spin coated to give thin transparent films. After corona poling and curing, the NLO properties were measured via Second Harmonic Generation (SHG) at 1064 nm. The coefficients d_{33} and r_{33} were calculated with the highest values of 37-45 pm/V and 9-10 pm/V, respectively.[6] However, heating the films to 180°C resulted in the reversible decay of SHG signal presumably from the loss of order due to randomization of chromophore orientation by rotation.[6]

Encouraged by this result, we reinvestigated the NLO properties of the *para*-nitroaniline material to develop a more efficient synthesis and to determine whether the temporal stability might be improved. Furthermore, although the $\mu_z\beta_z$ value for *para*-nitroaniline (PNA) chromophores is lower than nitrostilbene chromophores that are typically used as dopents in NLO systems (143 x 10^{-48} vs. 1150 x 10^{-48} esu),[7] chromophores such as 1 do have advantages. In addition to the higher thermal stability of PNA derivatives, highly conjugated chromophores absorb in the first and second harmonic of low power diode lasers (near 800 and 400 nm)[8] creating the possibility of resonance effects, low throughput, and optical damage.

EXPERIMENT

All reactions were conducted in oven dried glassware under an inert atmosphere (N_2 or Ar). Solvents were purified by standard procedures.[9] Characterization was performed as reported earlier.[10]

Monomer Synthesis

4-Nitro-N, N-bis(prop-3-enyl)-aniline (4). In a 250 mL flask equipped with a stir bar was added in order ethanol (100 mL), 1-fluoro-4-nitrobenzene (14.68 g, 0.10 mol), diallylamine (10.80 g, 0.11 mol, 1.07 eq) and triethylamine (10.16 g, 0.100 mol, 0.9 eq). The slightly yellow solution was heated to 80° C and stirred for 30 days. The solution was reduced in volume to ~20 mL *in vacuo*. This golden suspension was taken up in ether (100 mL) and washed with HCl (1 $\underline{M}$, 3 x 50 mL), NaHCO$_{3(aq)}$ (sat, 3 x 50 mL) and H_2O (2 x 50 mL). The water layers were combined and brought to pH ~7 by addition of NaOH (1 $\underline{M}$); residual product was extracted with ether (3 x 100 mL). The organic layers were combined, dried over MgSO$_4$ and filtered; volatile organics were removed *in vacuo* to give a deep yellow oil (21.5 g, 91.1 %). Flash chromatography gave an analytically pure yellow oil (19.3 g, 85.1 %). R$_f$ 0.30 (4:1 pet ether/ether); ^{1}H NMR (500 MHz, CDCl$_3$) δ 8.08 (AA'XX', J=7.34, 2.2 Hz, 2H, ArH), 6.10 (AA'XX', J=7.34, 2.2 Hz, 2H, ArH), 5.83 (m, 2H, NCH$_2$C$\underline{H}$=CH$_2$), 5.22 (dd, J=10.26, 1.1 Hz, 2H, NCH$_2$CH=C$\underline{H}_2$), 5.14 (dd, J=17.23, 1.1 Hz, 2H, NCH$_2$CH=C$\underline{H}_2$), 4.01 (dd, J=6.23, 1.46 Hz, 4H, NC$\underline{H}_2$CH=CH$_2$); ^{13}C NMR (70 MHz, CDCl$_3$) δ 153.2, 137.2, 131.7, 126.1, 116.9, 110.7, 52.9; IR (Neat) 752.1, 798.4, 825.4, 995.1, 1110.8, 1199.5, 1241.9, 1315.2 (ArNR$_2$), 1592.9 (ArNO$_2$), 1643.0, 2911.9, 3085.6 cm^{-1}; UV-VIS (c=2.7 x 1)$^{-5}$, cyclohexane) λ_{max} 354 (A = 0.44, ε = 16000), 226 (A = 0.18), 208 (A = 0.12) nm; MS (EI) *m/e* calc'd for C$_{12}$H$_{14}$N$_2$O$_2$: 218.1055, found 218.1054; LRMS 218, 191, 145, 130, 103, 77.

4-Nitro-N, N-bis[(3-triethoxysilyl)propyl]-aniline (1). A 250 mL flask was transferred into the glove box and H$_2$PtCl$_6$•nH$_2$O (0.031 g, 7.6 x 10^{-5} mol, 0.2 mol %) was added. The flask was removed from the glove box, and toluene (38 mL) and a stir bar were added. To the suspension was added N,N-diallyl-4-nitroaniline 5 (8.50 g, 0.039 mol). To the bright yellow solution was added triethoxysilane (16.30 mL, 14.58 g, 0.0887 mol, 2.27 eq) and after 15 minutes, a small amount of precipitate developed. The solution was slowly heated to 130° C. Upon reaching approximately 80° C (oil bath temperature) an exotherm became evident. Upon reaching 90° C, the solution changed from yellow to brown/black. Solution was heated to reflux and stirred for 24 hours at which time ^{1}H NMR indicated that the reaction was not complete (alkene resonances at 5.22, 5.14 and 4.17 ppm). After an additional 16 hours, the solution was cooled and filtered through Celite to remove residual Pt catalyst. Triethoxysilane and toluene were removed *in vacuo* to give analytically pure deep yellow oil (20.78 g, 0.0380 mol, 97.5 %): ^{1}H NMR (500 MHz, CDCl$_3$) δ 8.07 (d, J=9.53 Hz, 2H, ArH), 6.60 (d, J=9.53 Hz, 2H, ArH), 3.83 (q, J=6.96 Hz, 12H, Si(OC$\underline{H}_2$CH$_3$)$_3$), 3.36 (t, J=8.0 Hz, 4H, NC$\underline{H}_2$CH$_2$CH$_2$), 1.71 (m, 4H, NCH$_2$C$\underline{H}_2$CH$_2$), 1.23 (t, J=6.97 Hz, 18H, Si(OCH$_2$C$\underline{H}_3$)$_3$), 0.62 (t, J=8.0 Hz, 4H, NCH$_2$CH$_2$C$\underline{H}_2$); ^{13}C NMR (125 MHz, CDCl$_3$) δ 152.6, 136.2, 126.3, 110.0, 58.5, 53.4, 20.3, 18.3, 7.4; ^{29}Si NMR (99.9 MHz, CDCl$_3$) δ -46.2; IR (neat) 752.1, 790.7, 960.4, 1079.9, 1110.8, 1168.7, 1199.5, 1515.8, 1596.8, 2888.8, 2927.4, 2973.7 cm^{-1}; UV-VIS (c=1.5 x 10^{-5}, cyclohexane) λ$_{max}$ 234 (0.065), 370 (0.18, ε = 12000) nm; MS (CI, isobutane) *m/e* calc'd for C$_{24}$H$_{46}$N$_2$O$_8$Si$_2$: 546.2793, found 546.2802; LRMS 547, 355, 311, 163

Polysilsesquioxane Film Formation, Curing Profile, and SHG

Conditions developed by Oviatt[10] were used with modifications detailed in the text (*vide infra*). In general, ~1 mL of a 0.4 molar solution of **1** was prepared in dry *n*-butanol. To the yellow solution was added enough formic acid to provide 4.0 equivalents of water. The solution was agitated for 5-10 minutes and spun (2000 Hz, 30 s) onto indium-tin oxide coated quartz slides. The slide was placed in a dry nitrogen atmosphere, and the poling profile described in the **Results** was applied. Second harmonic generation experiments were performed at the Department of Chemistry, SUNY University at Buffalo, Buffalo, NY. Details for this experiment have been previously published.[11]

RESULTS

Our previously reported synthesis of **1** utilized nucleophilic aromatic substitution of 4-chloronitrobenzene with N, N-bis(3-triethoxysilylpropyl)amine.[10] Two main difficulties needed to be overcome in this synthesis. First, the synthesis of N, N-bis(3-triethoxysilylpropyl)amine was low yielding (3-35%) and required extensive purification.[10,12] Second, the aromatic substitution step required harsh conditions (150° C) and was low yielding (9%).[10, 13] Accordingly, we developed the following synthesis (Scheme 1).

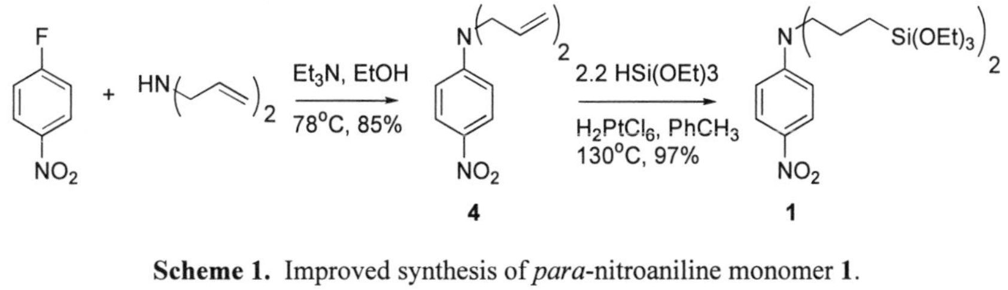

Scheme 1. Improved synthesis of *para*-nitroaniline monomer **1**.

The synthesis started by nucleophilic aromatic substitution of N,N-diallylamine and 4-fluoronitrobenzene.[14] However, we changed solvent from N-methyl pyrrolidinone to ethanol with triethylamine as the acid acceptor. Although our change in solvent from NMP to ethanol caused an increase in reaction time, it resulted in a higher yield (85 vs. 48%).[14]

Hydrosilylation of diene **2** with 2.2 equivalents of triethoxysilane with chloroplatinic acid as catalyst gave monomer **1**. After 40 hours at reflux, crude ^{1}H NMR indicated that the compound was primarily a γ silylated product as evidenced by only one triplet below 1 ppm. Removal of residual platinum catalyst by filtration through of Celite followed by removal toluene and triethoxysilane gave monomer **1** in quantitative yield. The high yield of **1** in comparison to low yields in hydrosilylations of secondary allyl amines[12] may lie in the electronic deficient nitrogen atom in **4** or from the decrease in triethoxysiliane.[15] The reduced electron density on the nitrogen may decrease the association of nitrogen with the platinum catalyst favoring the association of the Pt(0) with the alkene. Also, a reduction of triethoxysilane in the hydrosilylation reduces isomerization of the allylamine to the unreactive vinylamine.[12] This more efficient synthesis of N,N-(3-triethoxysilylpropyl)-4-nitroaniline is applicable to the preparation of large quantities (~10 g) in good yield (82.5%).

In our initial studies of polysilsesquioxane films (**x-1**) prepared with monomer **1** we used conditions developed earlier (0.4 M concentration, 4.0 eq. H$_2$O, formic acid).[13] After spin coating onto an ITO coated quartz slide, the film (**x-1a**) was poled for extended period (30 min vs. 4 hr) at 180°C. After this extended poling, the film was characterized by SHG. We then examined the SHG signal stability of at elevated temperatures (Figure 2).

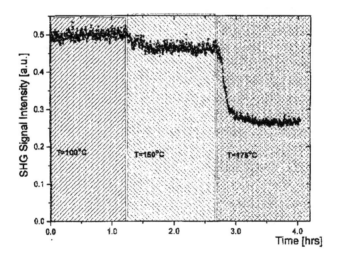

Figure 2. Thermal stability of poled alignment in thin-polysilsesquioxane films (**x-1a**).

Whereas the SHG of earlier materials prepared from **1** decayed at ~125°C,[6,13] these new materials (**x-1a**) showed no degradation after 1 hour at 100° C, retained 90% after 1.5 hours at 150°C, and retained 50% after 1.5 hours at 175°C.

Encouraged by these preliminary results, we examined more closely the thermal and temporal stability of the SHG signal of **x-1**. Although we did not change the conditions for the thin film formation of **x-1** (*vide supra*) we did change the poling and curing conditions. The poling field was increased from 5kV (~0.5 V/μm) to 135 V/μm. The earlier curing conditions

(180° C, air) were moderated (130° C, dry N_2) and the poling time was increased from 30 minutes to 2 hours (**x-1b**). The long-term thermal stability of the poled chromophore order of **x-1b** was probed by observation of the SHG signal at 100°C for 8 hours (Figure 3).

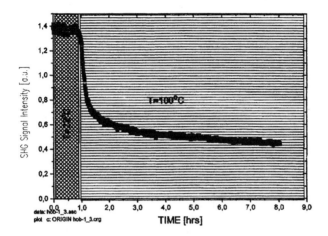

Figure 3. Stability of SHG signal from poled film of **x-1b** at 100° C.

After an initial decrease of 65%, the SHG signal stayed constant for 7 hours at 100° C. The long-term stability of poled order at ambient temperature was investigated by periodic examination of the SHG signal over 100 days (Figure 4).

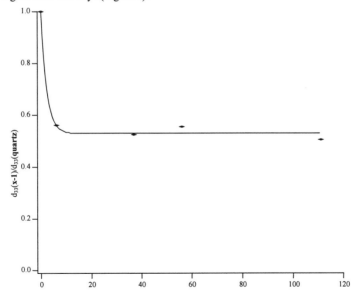

Figure 4. Temporal stability of imposed order in poled films of **x-1b**.

This normalized decay was modeled using the sum of two exponential terms (Equation 1).[16-18]

$$\frac{d_{33}(x-1)}{d_{33}(quartz)} = d_1 e^{\left(-t/\tau_1\right)} + d_2 e^{\left(-t/\tau_2\right)} \tag{1}$$

In the above model, τ_1 and τ_2 are short-term and long-term relaxation time constants, and d_1 and d_2 are constants that measure the relative importance of the two relaxation processes. Despite limited data, the model is appropriate (chi squared of 0.0013). The calculated coefficients compared favorably with a recently reported silicate/polysilsesquioxane containing a functionalized azo dye (Table I).[19]

Table I. Calculated coefficients for x-1 and polysilsesquioxane/silica thin films.

compound	d_1	τ_1 (days)	d_2	τ_2
x-1b	0.47	2.2	0.53	∞
50/50[a]	0.35	2.3	0.65	∞
30/70[a]	0.53	1.9	0.47	∞

[a]refers to the ratio of $Si(OMe)_4$ to bis-triethoxysilyl functionalized dye

Hence, the temporal stability of the dipole order in poled polysilsesquioxane thin films (x-1b) compares quite well with the stability of materials synthesized with an *external* crosslinker (TMOS). Due to the length of time necessary to obtain τ_2, it was assumed to be infinite.[18]

The calculation of $\chi^{(2)}$ was determined by angular dependence of the SHG signal. By comparing it to an Y-cut quartz crystal[20] a $\chi^{(2)}$ value of 9 x 10^{-8} esu was calculated.[21]

Thus, by increasing the intensity of the electric field and extending the duration of poling, both the temporal and thermal stability of the SHG signal from x-1a and x-1b was extended.

CONCLUSIONS

We have demonstrated that bridged polysilsesquioxanes synthesized from 4-nitro-N,N-bis[(3-triethoxysilyl)propyl]aniline 1 display second order NLO properties with a measured $\chi^{(2)}$ value of 9 x 10^{-8} esu, a calculated d_{33} coefficient of 18.9 pm/V, and an r_{33} value of 4.7 pm/V. The temporal stability of the long-range order necessary for second order NLO effects was improved by performing the poling in a dry nitrogen atmosphere, increasing the intensity of the poling field, changing the curing temperature, and increasing the curing time. By using a bis-exponential function we were able to model the temporal stability of the chromophore. The coefficients that describe the temporal stability (τ_1 and τ_2) compare quite well with materials prepared with external crosslinkers.

ACKNOWLEDGMENTS

We acknowledge the Air Force Office of Scientific Research for funding. S. T. Hobson gratefully acknowledges the United States Army Research Institute for Chemical Defense for encouraging the publication of this work and financially supporting the presentation of this data.

REFERENCES

* The opinions or assertions contained herein are the private views of the authors and are not to be contrued as official or as reflecting the views of the Army or the Department of Defense.

1. Prasad, P. N. in *Contemporary Nonlinear Optics* Agrawal, G. P.; Boyd, R. W.; Eds. Quantum Electronics--Principles and Applications, Academic Press, Boston, 1992; Chapter 7.

2. Zyss, J.; Chemla, D. S. in *Nonlinear Optical Properties of Organic Molecules and Crystals* Chemla, D. S.; Zyss, J.; Eds. Academic Press, New York, 1987; 23
.

3. a Burzynski, R.; Prasad, P. N. in *Sol Gel Optics: Processing and Applications*; Klein, L. C., Ed.; Kluwar Academic Publishers, 1994; pp. 417-450. b Levy, D., Esquivia, L. *Adv. Mater.* **1995**, *7*, 120.

4. Peng, Z.; Yu, L. *Macromolecules* **1994**, *27*, 2638

5. Meyeriux, R.; Lemomte, J.; Tapolsky, G. *Proc. SPIE* **1991**, *1560*, 454

6. Oviatt, Jr., H. W.; Shea, K. J.; Lairo. S.; Shi, T.; Dalton, Steier, W. H.; Dalton, L. R. *Chem. Mater.* **1995**, *7*, 493.

7. Measured by Electric Field Induced Second Harmonic Generation EFISH, see Ulman, A.; Willand, C. S.; Kohler, W.; Robello, D. R.; Williams, D.J.; Handley, L. *J. Am. Chem. Soc.* **1990,** *112*, 7083.

8. Tam, W.; Cheng, L. l Bierlein, J. D.; Cheng, L. K.; Wang, Y.; Feiring, A. E.; Meredith, G. R.; Eaton, D. F.; Calabrese, J. C.; Rikken, G. L. J. A. in *Materials for Nonlinear Optics : Chemical Perspectives* ; Marder S. R.; Sohn J. E., editor ; Stucky G. D. eds. ACS symposium series 455; American Chemical Society: Washington, DC, 1991; pp. 158-160.

9. Perrin, D. D., Armarega, W. L. F. *Purification of Laboratory Chemicals*; Pergamon Press: Oxford, 1988.

10. Hobson, S. T. and Shea, K. J. *Chem. Mater.* **1997**, *9*, 616.

11. Mortazavi, M. A.; Knoesen, A.l Kowel, S. T.; Higgins, B. G.; Dienes, A. *J. Opt. Soc. Am. B.* **1988**, *6*, 733.

12. Small, J. H. Ph.D. Dissertation, University of California, Irvine, 1995, p 139.

13. Oviatt, H. W. Ph.D. Dissertation, University of California, Irvine, 1995, Ch 8.

14. Gibbons, W. M.; Grasson, M. K.; O'Brien, M. K.; Shannon, P. J.; Sun, S. T. *Macromolecules* **1994**, *27*, 771.

15. Marciniec, B.; Gulinski, J.; Urbaniak, W.; Kornetka, Z. *Comprehensive Handbook on Hydrosilylation,* Marciniec, B., Ed. Pergamon, Oxford.

16. Hampsch, J. M.; Torkelson, J. M.; Bethke, S. J.; Grubb, S. G. *J. Appl. Phys.* **1990**, *67*, 1037.

17. Stahelin, M.; Walsh, C. A.; Burland, D. M.; Miller, R. D.; Twieg, R. J.; Volksen, W. *J. Appl. Phys.* **1993**, *73*, 8471.

18. Suziki, A.; Matsuoka, Y. . *J. Appl. Phys.* **1995**, *77*, 965.

19. Lebeau, B.; Bresselet, S.; Zyss, J.; Sanchez, C. *Chem. Mater.* **1997**, *9*, 1012.

20. d11 for quartz = 0.81 x 10-9 esu.

21. Saleh, G. E. A.; Teich, M. C. *Fundamentals of Photonics;* John Wiley and Sons, Inc.: New York, 1991; p 780.

SECOND ORDER NONLINEAR OPTICAL THIN FILMS FABRICATED FROM ELECTRIC FIELD-ASSISTED ELECTROSTATIC SELF-ASSEMBLED MONOLAYER METHOD

Y. Liu *, R.O. Claus **, D. Marciu ***, C. Figura ***, and J.R. Heflin ***
*NanoSonic, Inc., 200 Country Club Drive, A-2, Blacksburg, VA 24060, nano@usit.net
**Fiber & Electro-Optics Research Center, Virginia Tech, Blacksburg, VA 24061-0356
***Department of Physics, Virginia Tech, Blacksburg, VA 24061-0435

ABSTRACT

A new method for the build-up of non-centrosymmetric multilayer thin films has been developed for the first time using an electric field-assisted electrostatic self-assembled monolayer (EF-ESAM) technique. An increase by 116% of the second-harmonic intensity of the films has been observed in comparison with that of ESAM film.

INTRODUCTION

We have recently demonstrated that ESAM processing inherently yields noncentrosymmetric molecular structures that possess large second order NLO response, without the need for electric filed poling [1-4]. The development of ESAM $\chi^{(2)}$ thin films provides significant advantages over the fabrication of organic $\chi^{(2)}$ thin films by alternative methods. For example, ESAM films can exhibit long-term stability of $\chi^{(2)}$ in contrast to electric field poling of glassy polymer [5], can provide thicker (tens of microns in thickness) films than the Langmuir-Blodgett technique, can be fabricated much more rapidly than covalent self-assembly [4] methods. Moreover, ESAM $\chi^{(2)}$ films has demonstrated remarkable thermal stability over 18 months after many cycles of heating up to 150 ^{0}C. Up to date, we have designed 14 ionic polymers optimized specifically for compatibility with the ESAM process and have fabricated many NLO films hundreds of bilayer thickness with consistent, reproducible film growth with each bilayer.

Here, we show that it is possible to modify the molecular orientation of ESAM films by applying an electric field during the ESAM deposition process. The resulting so-called EF-ESAM films exhibit enhanced $\chi^{(2)}$ values with an increase of a factor of two.

EXPERIMENT

The ESAM technique for the deposition of NLO films has been reported in detail before [4]. The EF-ESAM films are grown monolayer by monolayer by first immersing a pre-cleaned ITO-coated glass substrate in the poly(allylamine hydrochloride) (PAH) solution for 20 seconds. This is followed by rinsing and then by immersion of the substrate and an ancillary electrode parallel to each other into a polymer dye (poly S-119, Sigma) solution for 5 seconds. The electrodes were separated about 10 mm, and a positive voltage bias in the range of 0-1 V was applied to the ITO substrate, it has been demonstrated that a full monolayer coverage of a EF-ESAM film takes less than 50 microseconds by monitoring the deposition process [6]. This dipping process can be repeated as many times as desired until a film with the chosen number of bilayers has been produced.

The second-harmonic generation (SHG) experiments were carried out using both the 1064 nm fundamental wavelength of a Q-switched Nd:YAG laser and the 1200 nm output from a broadband, BBO optical parametric oscillator (OPO). The OPO was pumped by the 355 nm third harmonic of the Nd:YAG and was continuously tunable from 400 to 2500 nm. The incident intensity and polarization on the sample were controlled by a pair of polarization prisms. The film was rotated 45^0 away from normal incidence about the vertical axis, and the incident light was p polarized.

RESULTS AND DISCUSSIONS

The EF-ESAM films were first characterized by a Hitachi U-2010 UV-vis spectrometer. Figure 1 shows the UV-vis absorbance of the EF-ESAM Poly S-119/PAH film with 30 bilayers deposited on ITO substrate, in comparison, the optical absorbance of the ESAM Poly S-119/PAH having 68 bilayers is also shown in Figure 1.

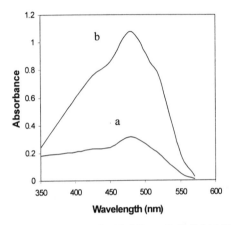

Figure 1. Optical absorbance spectra of a 30- bilayer Poly S-119 EF-ESAM film (a) and a 68-bilayer Poly S-119 ESAM film (b).

The film is orange in color and exhibits exceptional homogeneity and lack of scattering. Since the optical absorbance depends on the amount of materials deposited, we can infer that the amount of materials adsorbed in a 30-bilayer EF-ESAM film is equal to that of a 20-bilayer ESAM film.

Figure 2 illustrates the dependence of the SHG signal intensity as a function of the incident fundamental intensity for a single-sided, 30 bilayer Poly S-119 EF-ESAM film and the 68 bilayer Poly S-119 ESAM film with a $\chi^{(2)}$ value of 0.67 x 10^{-9} esu now serving as a standard reference sample. The data of the 30-bilayer EF-ESAM film are seen to be in excellent agreement with the quadratic dependence. This demonstrates that the orientation of the chromophores is the same for each successive layer. If the degree of orientation was decreased for the latter deposited layers, the SHG intensity would yield a subquadratic dependence on the number of bilayers. The $\chi^{(2)}$ of the 30-bilayer EF-ESAM film can be rapidly determined by comparison to this reference.

Since the thickness of the EF-ESAM film is much less than the second-harmonic coherence length, the observed $\chi^{(2)}$ intensity increases quadratically with the thickness of a given material.

$$\chi^{(2)} = \chi^{(2)}_{ref}\sqrt{\frac{A}{A_{ref}}}\frac{l_{ref}}{l} \qquad (1)$$

30

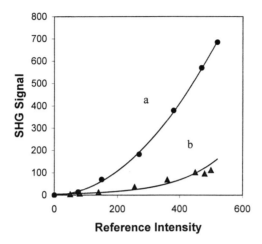

Figure 2. Comparison of second-harmonic intensities observed for a 30-bilayer Poly S-119 EF-ESAM film (a) and a 68-bilayer Poly S-119 ESAM film (b).

In this limit, the $\chi^{(2)}$ value is determined by Equation 1, where A is the coefficient of the fit to the second-harmonic intensity as a function of fundamental intensity and l is the optical path length through the sample, i.e., the film thickness.

From Figure 2, it can be seen that the second-harmonic intensity $\chi^{(2)}$ is significantly larger than that of the reference film, although the reference sample is 3 times thicker than the EF-ESAM film. From Equation 1, a $\chi^{(2)}$ value of 1.45 x 10^{-9} esu, or a 116% increase, is obtained. A possible suggestion for this enhancement of SHG is that the electric field affects the alignment of the polymer dye molecules, which results in a preferred dipole orientation for SHG effects. The detailed study of the influence of electric field on the layer-by-layer deposition is being carried out in detail using polarization modulation infrared reflection absorption spectroscopy and electrochemistry techniques.

CONCLUSIONS

In this paper, we have shown that films deposited using the technique of electric field assisted electrostatic self-assembled monolayer (EF-ESAM) processing can spontaneously assemble into a noncentrosymmetric structure with a high $\chi^{(2)}$ value. The amount of material adsorbed in each layer of the EF-ESAM films is significantly less than that in ESAM films without the external field. Our results have demonstrated that the dipole orientation of the ESAM NLO films can be influenced significantly during the deposition process under the external electric field, and an increase of a factor of 2 of the $\chi^{(2)}$ value has been achieved.

ACKNOWLEDGEMENTS

This work was supported by Air Force contract F33615-98-C-5421, Office of Naval Research contract N00014-96-1-1285, and National Science Foundation grants DMI-9860084.

REFERENCES

1.J. R. Heflin, Y. Liu, C. Figura, D. Marciu, and R.O. Claus, SPIE, **3147**, 10 (1997).
2. J. R. Heflin, Y. Liu, C. Figura, D. Marciu, and R.O. Claus, in *Organic Thin Films for Photonic Applications*, Vol. 14, OSA Technical Digest Series, Optical Society of America, Washington, DC 1997, P. 78-80.
3. K. M. Lenahan, Y. Wang, Y. Liu, R. O. Claus, J. R. Heflin, D. Marciu, and C. Figura, Adv. Mater. **10**, 853 (1998).
4. , J. R. Heflin, C. Figura, D. Marciu, Y. Liu and R. O. Claus, Appl. Phys. Lett. **74**, 495 (1999).
5. K. D. Singer, S. J. Lalama, and J. E. Sohn, Appl. Phys. Lett. **49**, 248 (1986).
6. W. Zhao, Y. Liu, Y. Wang, and R. O. Claus, in preparation.

NONLINEAR OPTICAL FILMS BY ALTERNATING POLYELECTROLYTE DEPOSITION ON HYDROPHOBIC AND HYDROPHILIC SUBSTRATES

M. JOSEPH ROBERTS

[a]United States Department of the Navy, Naval Air Warfare Center Weapons Division, Materials and Chemistry Division, Code 4T4220D, 1 Administration Circle, China Lake, CA 93555

ABSTRACT

The formation of acentric films using alternating polyelectrolyte deposition (APD) has been achieved on hydrophobic and hydrophilic glass substrates. APD is a layer-by-layer technique for the formation of polymer films by alternately immersing a substrate in aqueous solutions of a polyanion and a polycation. APD provides precise control of the overall film thickness that through automated processing may exceed a thousand layers. In this study, APD films were made of an NLO-active polycation, stilbazolium-substituted polyepichlorohydrin (SPECH), and NLO-inactive polyanions. The peak maximum UV-Visible absorbance in transmission through the films was linear as a function of the number of bilayers. Second harmonic generation (SHG) was used as a tool to indicate acentric order of polarizable side-chain chromophores within the APD films. The SHG exhibited the expected quadratic intensity increase with film thickness.

INTRODUCTION

Alternating polyelectrolyte deposition (APD) is a room temperature self-assembly technique for the formation of multilayer stacks without the need for expensive equipment.[1] A multilayer film is formed by alternately immersing the substrate in aqueous solutions of a polyanion and a polycation. Since the thickness of each layer is on the order of 2 nanometers, APD allows precise control of the overall film thickness. APD films of thousands of layers have been reported. Coulombic forces are responsible for the formation of the APD films so the resulting structure is very robust. Polishing is necessary for removal of an APD film from the substrate. The technique has been used to produce films incorporating a range of materials including electrically conducting polymers,[2] conjugated polymers for light emitting diodes (LED),[3] proteins,[4] nanoparticles,[5] and noncentrosymmetric-ordered second order nonlinear optical polymers (NLOPs).[6]

Noncentrosymmetric chromophore orientation has been observed in APD films by several research groups confirming that polar ordered films are obtained.[7-9] Uniform NLO APD films have been deposited with SHG increasing quadratically with film thickness over one hundred bilayers, although $\chi^{(2)}$ values were modest and comparable to quartz.[10-11] More recently, our group has made NLO APD films in which both the polyanion and the polycation contain polarizable chromophores and the results suggest that both polyelectrolytes contribute to the observed SHG signal.[12] Also, our group has demonstrated that the chromophore orientation in NLO APD films is stable at temperatures up to 150 °C.[13]

Recent efforts have focussed on a polyepichlorohydrin substituted with a side-chain stilbazolium chromophore (SPECH) (Figure 1). This paper is a report our recent progress with the APD of SPECH with polystyrene sulfonate and SPECH with polyacrylic acid.

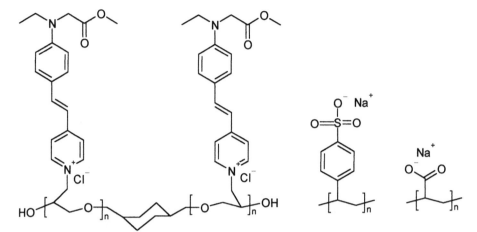

Figure 1. NLO-active water-soluble polycation, SPECH, (left), inactive water-soluble polyanions, PSS, (middle), and PAA, (right) used to make polar films by APD. For SPECH, n=4-5; for PSS, n=300-370; and for PAA, n=750-800.

EXPERIMENTAL

The synthesis of SPECH will be reported in another publication.[14] All polymer solutions were made from ultrapure water taken from a Barnstead Nanopure purification system (17.9 MΩ-cm resistivity, 0.2 micron filter). The solutions were made by dissolving SPECH to give a concentration of 10^{-6} moles of ionic repeat units per liter. The SPECH solutions were filtered through 0.2-micron pore-sized membrane Millipore filters directly into the dipping vessels. Although slow hydrolysis of the stilbazole is possible, there was no evidence of hydrolysis: the peak absorbance and wavelength of the SPECH solutions remained constant for over 14 hours. At no time were other salts added. Poly(sodium 4-styrenesulfonate), PSS, having a weight-average molecular weight of 70,000 (Aldrich, Milwaukee, WI) was used to make 10^{-4} M solutions by diluting the as-received 20 weight percent water solution with ultrapure water. Poly(acrylic acid), PAA, having a weight-average molecular weight of 50,000 (Polysciences, Warrington, PA) was used to make 10^{-4} M solutions by diluting the as-received 25 weight percent water solution with ultrapure water. The pH of the PAA solution was adjusted by adding dilute aqueous solution of NaOH until the desired pH of 6.5 was reached as measured by a pH meter. Solutions were stored in the dark not more than a few hours at approximately 23°C before adding to the dipping vessels. The dipping vessels were plastic baths used in a Zeiss HMS Programmable Slide Stainer (Carl Zeiss, Thornwood, NY). The substrates were glass microscope slides (Fisher Scientific, Catalog # 12-550A). The glass slides were cleaned with $H_2SO_4/H_2O_2/H_2O$ (7:1:2) for 60 minutes, rinsed with ultrapure water, and dried at 105 °C for 1 to 2 hours. Cleaned slides were made hydrophobic by exposure to neat hexamethyldisilazane vapor for over an hour at room temperature. The surface energy of the hydrophobic glass slides was characterized by contact-angle measurements. A contact angle of 90° is typical for complete trimethyl silyl coverage of glass. The contact angle of these substrates was about 88°. The hydrophobic glass slides were used immediately after rinsing with ultrapure water to remove adhering particles. To make hydrophilic glass slides, the acid cleaning procedure above was followed by rinsing with ultrapure water, cleaning with $NH_4OH/H_2O_2/H_2O$ for 10 minutes, and drying.

The solutions were kept at room temperature (approximately 23°C) during the film depositions. Depositions were performed in a Class 100 clean room under filtered (blue through ultraviolet blocked) fluorescent lighting. The APD of SPECH with PSS on hydrophobic substrates has been described previously.[13] For the deposition of SPECH on hydrophilic glass, the slide was immersed in the polycation solution for 5 minutes, which was just long enough to develop the maximum SH intensity of the first SPECH layer. The slide was rinsed 3 times in ultra pure water for 10 seconds. For the deposition of PAA, the glass slide containing a SPECH polymer layer on the surface was immersed in a 10^{-4} M PAA solution for 5 minutes. The slide was removed, rinsed in ultra pure water 3 times for 10 seconds. Subsequent bilayers were built up on the substrate by alternating between the polycation and polyanion solutions with the same rinsing steps after each deposition.

The characterization of the films by UV-Visible spectroscopy and SHG measurements has been described previously.[13] To the naked eye, the NLO APD films were featureless, highly uniform, and transparent with a shiny surface.

RESULTS AND DISCUSSION

The SPECH is a low molecular weight polyelectrolyte and so most likely adopts a chain-extended conformation. However, the side chain chromophores of the SPECH represent the greatest fraction of the total mass of any complete SPECH molecule. The stilbazolium side-chain chromophores in SPECH possess a strong ground state dipole moment. Adjacent stilbazolium chromophores will electrostatically repel one another yet they are covalently bound to the main chain so the net result is a 'bottle-brush' conformation of the SPECH in dilute aqueous solution. The SPECH is expected to adsorb to the surface in an extended conformation.

The length of the PSS chain is greater than its estimated persistence length (15 nm) and so is expected to be an expanded random coil in solution. Computational molecular modeling calculated distance between charges along the chain in the extended conformation is approximately 0.62 nm for both SPECH and PSS. The distance between charges for PAA was not calculated. The adsorption of the first layers of SPECH and PSS on a hydrophobic substrate has been figuratively described previously.[9]

Adsorption of the first layer of SPECH on a hydrophilic substrate

The SPECH adsorbs rapidly from solution onto base-washed glass. Adsorption of the first layer is complete within 5 minutes of immersion. Also the adsorbed SPECH exhibits relatively strong polar order within the same short period of time. Apparently, the positive charges of the pyridinium end of the chromophores in SPECH quickly forms ion pairs with the O⁻ sites at the surface of base-washed glass. The process is facilitated by the extended conformation of the SPECH in solution plus the chromophores may easily tilt up away from the plane of the substrate to accommodate the later adsorbing SPECH chains.

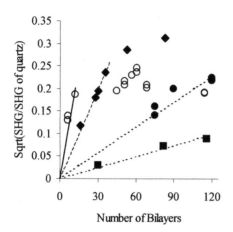

Figure 2. SHG intensity versus the number of SPECH/PSS bilayers deposited on hydrophobic glass (filled symbols) and SPECH/PAA on hydrophilic glass (open circles). Data for SPECH/PSS films in which the adsorption time of SPECH was 30 minutes (filled squares), 40 minutes (filled circles), and 50 minutes (filled diamonds) is shown.

Adsorption of the first layer of PAA on the first layer of SPECH on a hydrophilic substrate

At pH 6.5, most of the carboxylic acid groups are converted to the sodium salt and so each PAA polymer chain is highly charged. PAA adsorbed from pH 6.5 solutions has been shown to form layers less than 0.3 nm thick with minimal RMS roughness.[17] In contrast to the adsorption of PSS onto SPECH layers on hydrophilic glass which disrupts the polar order of the SPECH layer,[9] PAA does not disturb the SPECH layer to any measurable extent. This allowed the formation of multilayers of SPECH and PAA with quadratic increase of SHG and linear increase of UV-visible absorbance as the film thickness increased up to 10 bilayers. Further deposition resulted in continued linear increase in UV-Visible absorbance but the SHG intensity did not continue to increase quadratically (Figure 2).

Although the deposition process is slower, uniform SPECH/PSS films up to 120 bilayers thick have been made. The UV-visible absorbance in transmission through the films is linear as a function of the number of deposited bilayers. The SHG intensity follows the expected quadratic dependence when the immersion time for each layer of SPECH is 30 or 40 minutes (Figure 2). At immersion times less than 30 minutes the SHG signal was below the detection limit.

CONCLUSIONS

In this work, thin films with polar ordered chromophores were obtained by APD. The best results were obtained alternating SPECH with PSS on hydrophobic substrates and SPECH with PAA on hydrophilic substrates. Future work will include further characterization of the APD films by Kelvin probe microscopy, ellipsometry, and azimuthal SHG measurements.

Design and synthesis of new polymer systems is underway. Work is now focussed on improving the $\chi^{(2)}$ of the APD films by extending the uniform deposition of the SPECH/PAA to greater film thickness.

ACKNOWLEDGMENTS

The ONR Division of Physical Sciences provided funding for this work.

REFERENCES

1) G. Decher, Y. Lvov, J. Schmitt, Thin Solid Films **244**, 772 (1994).

2) C. N. Sayre, D. M. Collard, Journal of Materials Chemistry **7**(6), 909 (1997).

3) J. H. Ferreira, M. Cheung, M. Rubner, Thin Solid Films **244**, 806 (1994).

4) Y. Lvov, K. Ariga, I. Ichinose, T. Kunitake, Thin Solid Films **284-285**, 797 (1996).

5) J. Schmitt, G. Decher, W. J. Dressick, S. L. Brandow, R. E. Geer, R. Shashidar, J. M. Calvert, Advanced Materials **9**(1), 61 (1997).

6) Y. Lvov, S. Yamada, T. Kunitake, Thin Solid Films **300**, 107 (1997).

7) X. Wang, S. Balasubramanian, L. Li, X. Jiang, D. J. Sandman, M. F. Rubner, J. Kumar, S. K. Tripathy, Macromol. Rapid Commun. **18**, 451 (1997).

8) A. Delcorte, P. Bertrand, E. Wischerhoff, A. Laschewsky, Langmuir **13**, 5125 (1997).

9) M. J. Roberts, J. D. Stenger-Smith, P. Zarras, G. A. Lindsay, R. A. Hollins, A. P. Chafin, R. Y. Yee, K. J. Wynne, Proceedings of the SPIE **3281**, 126 (1998).

10) J. R. Heflin, Y. Liu, C. Figura, D. Marciu, R. O. Claus, Technical Digest of the OSA/ACS Conference on Organic Thin Films for Photonics Applications, 78 (1997).

11) K. M. Lenahan, Y. Wang, Y. Liu, R. O. Claus, J. R. Heflin, D. Marciu, C. Figura, Advanced Materials **10**(11), 853, (1998).

12) G. A. Lindsay, M. J. Roberts, J. D. Stenger-Smith, P. Zarras, R. A. Hollins, A. P. Chafin, R. Y. Yee, K. J. Wynne, Proceedings of the SPIE **3474**, 63 (1998).

13) M. J. Roberts; W. N. Herman; G. A. Lindsay; K. J. Wynne, Journal of the American Chemical Society **120**, 11202 (1998).

14) G. A. Lindsay, M. J. Roberts, J. D. Stenger-Smith, P. Zarras, R. A. Hollins, A. P. Chafin, R. Y. Yee, K. J. Wynne, Journal of Materials Chemistry, in press.

15) J.J. Tanahatoe, M. E. Kuil, Macromolecules **30**, 6102 (1997).

16) V. V. Tsukruk, V. N. Bliznyuk, D. Visser, A. L. Campbell, T. J. Bunning, W. W. Adams, Macromolecules **30**, 6615.

17) S. S. Shiratori, M. F. Rubner, Technical Report of the Institute of Electronics, Information, and Communications Engineers OME98-106, 32, (1998).

STRUCTURE-PROPERTY RELATIONSHIPS IN
ORGANIC NONLINEAR OPTICAL MATERIALS

ERIC M. BREITUNG, ROBERT J. MCMAHON *
Department of Chemistry, University of Wisconsin, Madison, WI 53706-1369.
breitung@students.wisc.edu, mcmahon@chem.wisc.edu

ABSTRACT

Tuning the degree of bond-length alternation in organic nonlinear optical materials is a powerful paradigm for the design of organic materials with large molecular hyperpolarizabilities (β). Several research groups have employed this paradigm in the design and synthesis of NLO materials incorporating donor-acceptor polyenes. Increased bond-length alternation in polyenes leads to decreased barriers to rotation about C=C bonds and, hence, increased conformational flexibility. Since the degree of bond-length alternation is solvent dependent, so is the degree of conformational flexibility. In an effort to probe the influence of conformational flexibility on NLO response, we synthesized a series of simple donor-acceptor polyenes that are either conformationally flexible (**1a**, **2a**, **3a**) or rigid (**1b**, **2b**, **3b**). For each pair of molecules ZINDO sum-over-states calculations predict a larger value of β_μ for the conformationally flexible isomer, but EFISH measurements ($CHCl_3$) display mixed results. Various explanations for this behavior will be considered.

INTRODUCTION

Significant interest exists for the design and development of materials exhibiting large second-order nonlinear optical properties due to their potential application in telecommunications, optical computing, and optical signal processing.[1,2] Optimizing both the thermal stability and magnitude of second-harmonic generation has been the focus of NLO research for the past decade.[3,4] Much of this work involved the use of organic chromophores.

Typical organic NLO compounds include donor (D) and acceptor (A) moieties connected by a π-conjugated linker. A common recent strategy for optimizing second-order nonlinear optical properties has involved varying the degree of bond length alternation (BLA) of the π-conjugated bridge.[5,6] BLA is defined as the difference of the average carbon-carbon single and double bond lengths of the unit bridging the donor and acceptor. The BLA can be related to the contributions of two limiting resonance structures to the overall ground state electronic structure. The limiting structures refer to polyenic and zwitterionic resonance forms (Scheme 1). Complete contribution from either the polyene or zwitterion form to the ground state leads to severe bond length alternation (0.1 to -0.1 Å, respectively), which correlates to minimal electronic mixing between the donor and acceptor. Partial contribution from both resonance forms leads to a

polyene   zwitterion   (Scheme 1)

diminished |BLA|, relating to an increase in donor/acceptor mixing. Increasing the mixing between donor and acceptor or, equivalently, decreasing the |BLA| from 0.1 to 0.05 Å has been shown to increase the hyperpolarizability, $|\beta_\mu|$. Further mixing to the point where the polyene and zwitterion forms contribute equally (BLA = 0 Å) can lead to vanishingly small hyperpolarizabilities.

Computational studies predict that the presence of a small electric field about a D-π-A chromophore can influence the BLA parameter, which can dramatically effect β_μ.[6-8] A simple method of experimentally altering the electric field to which a molecule is exposed is to change the solvent polarity used to dissolve the NLO chromophore. It is well known, however, that altering the solvent polarity can decrease the barrier to rotation about carbon-carbon double bonds for donor-acceptor substituted polyenes.[9,10] This increase in the conformational flexibility should also significantly effect the NLO properties. In an effort to probe the influence of conformational flexibility on NLO response, we synthesized a series of simple donor-acceptor polyenes that are either conformationally flexible 1a-3a or rigid 1b-3b (Scheme 2).

1a

1b

2a

(Scheme 2)

2b

3a

3b

In this paper we report the computed and experimentally determined absorption spectra and NLO properties as well as the calculated dipole moments of a series of conformationally rigid and flexible chromophores. Semi-empirical computations provide structural data with which trends found in the experimental and computational hyperpolarizability measurements will be explained.

RESULTS

Table 1 contains the computed[11] and experimentally determined absorption spectra and NLO properties for compounds **1-3**. The EFISH experiments show no consistent trend differentiating the conformationally rigid chromophores from their respective flexible analogs. For compounds **1-2** the rigid and flexible analogs display approximately equal β_μ values. Compounds **3a** and **3b**, however, display significant and different hyperpolarizabilities. The flexible and rigid pairs with small hyperpolarizabilities (**1-2**) may display similar hyperpolarizabilities because of the large error inherent to the EFISH experiment (10-15%).

Semi-empirical ZINDO calculations predict larger hyperpolarizabilities for the flexible compounds relative to their respective rigid chromophores for all of the compounds studied. The ZINDO calculations do not, however, consistently predict the absolute magnitudes of the molecular hyperpolarizability. The calculated β_μ values for **2b**, **3a**, and **3b** are significantly lower than the experimentally determined hyperpolarizabilities.

To explain the trends displayed in the ZINDO calculations, we concider conformational aspects of the AM1 optimized geometries.[12] The dihedral angle defined by LP-N1-C3-C4 (ϕ_a) (Scheme 3) and the BLA parameter are reported in Table 2. The dihedral angles of the bridging

Table 1. Electronic Absorption, Dipole Moment, and Nonlinear Optical Properties for Compounds 1-3.

Compound	λ_{max} [a,d] calc	λ_{max} [a,g] expt	μ [c,d] calc	$\mu \cdot \beta$ [d,e] calc	$\mu \cdot \beta$ [b,e] expt	β_μ [d,f] calc	β_μ [b,f] expt
1a	271		7.1	29	7	4	1
1b	266	233	6.9	14	13	2	2
2a	311	331	8.2	98	78	12	10
2b	396	339	8.3	50	90	6	11
3a	409	381	8.8	230	408	26	46
3b	394	386	9.5	124	311	13	33

[a] in units of nm; [b] in $CHCl_3$; [c] in units of Debye (1 D = 10^{-18} cm esu); [d] computed from AM1 geometry; [e] in units of $10^{-48}cm^6$; [f] in units of 10^{-30} cm^5 esu^{-1}, [g] in hexane.

polyene chains are computed to be in a virtually planar s-trans configuration. The π-system of the acceptors are predicted to be co-planar with the bridging polyene moieties. Examination of the dihedral angles provides an explanation for the observed disparities in NLO response between rigid and non-rigid chromophores with identical π-systems.

DISCUSSION

For all compounds, computationally predicted (ZINDO) hyperpolarizabilities are greater for the flexible relative to the rigid chromophores. Experiment corroborates these calculations for only compounds **3a** and **3b**, whereas compounds **1-2** may not follow the predicted trend due to the error associated with the EFISH experiment. Differences in the hyperpolarizability between the respective rigid and flexible pairs may arise from the ability of the flexible isomers to achieve conformations other than those shown in Scheme 2 by Z/E isomerization of the carbon-carbon double bonds and/or rotating about the carbon-carbon single bonds. Strong evidence exists for the ability of donor-acceptor polyenes to readily isomerize about the carbon-carbon double bonds.[9,10] Hence, the flexible isomers can potentially rotate about the bridging bonds. If isomerization about single or double carbon-carbon bonds was occurring in the flexible chromophores, a decreased hyperpolarizability would be expected for those chromophores relative to the rigid compounds. Because this is the opposite of the observed trend for **3a** and **3b**, it is likely not the case, at least in $CHCl_3$ that conformational issues regarding isomerization about the bridging carbon-carbon bonds can account for the lower hyperpolarizabilities observed for the rigid chromophores.

A significant literature exists in which the BLA parameter is used to explain changes in the hyperpolarizability. Changes in the local electric field surrounding the chromophore (solvent polarity) caused variations in β_μ, which were explained through changes in the electronic nature of the bridging bonds via the BLA parameter.[5,13] The rigid and flexible pairs in this study have identical donors, acceptors, and π-systems. Because of this, assuming no single or double bond isomerization, the rigid and flexible pairs are expected to display similar bond length alternation and, hence, similar β_μ values. Examination of Table 2 shows that all of the compounds in this study are predicted to have approximately the same BLA parameter (ca. 0.10 Å), so differences in BLA can not account for the differences observed in β_μ.

Table 2. Computed Dihedral Angles and Bond Length Alternation Values for Compounds 1-3.

Compound	$\phi_a^{a,c}$	$BLA^{b,c}$
1a	82.9	0.097
1b	68.5	0.097
2a	81.7	0.097
2b	66.9	0.098
3a	81.2	0.094
3b	66.1	0.098

[a] in degrees; [b] in Å; [c] computed from AM1 geometry

Scheme 3

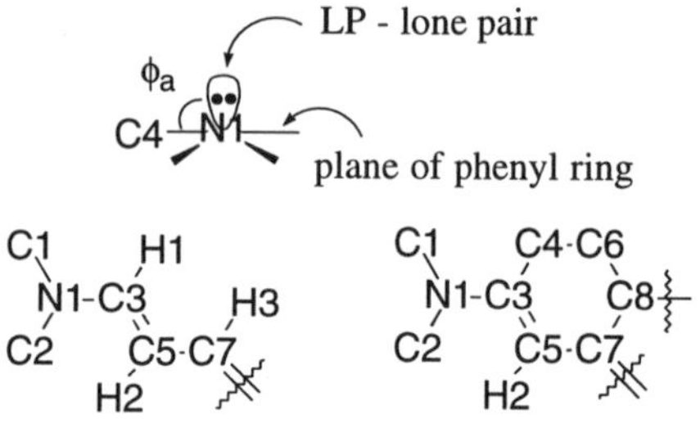

The main structural difference between the rigid and flexible pairs is the presence of the fused cyclohexene rings used to lock the conformation of the rigid analogs. Neither severe twisting about the single or double carbon-carbon bonds of the bridge nor BLA differences can account for the observed changes in β_μ. Examination of other conformational issues leading to electronic differences is appropriate. The AM1 optimized geometries used as input to the ZINDO calculations of β_μ provide useful conformational information.

Analysis of the dihedral angle ϕ_a (LP-N1-C3-C4) (Scheme 3, Table 2) describes the degree to which the lone pair (LP) of the amine nitrogen is displaced from the conjugated p-orbitals of the bridging system. The p-orbital overlap displays a $\sin^2\phi_a$ dependence, whereby small changes in ϕ_a translate into large changes in computed hyperpolarizability. For the greatest amount of overlap of the lone pair with the p-orbital of C3, ϕ_a should be 90°. For compounds 1a, 2a, and 3a, ϕ_a ranges from 81.2 - 82.9°, whereas the rigid analogs 1b, 2b, and 3b range from 68.5 - 66.1°. The differences in p-orbital overlap of the amine lone pair with the bridging carbon p-orbitals between the rigid and flexible pairs is approximately 15%. The loss of overlap predicted for the rigid chromophores is likely the source of the decreased calculated β_μ values for 1b, 2b, and 3b relative to 1a, 2a, and 3a. The decreased ϕ_a from compounds 1b, 2b, and 3b is due to the steric repulsions between the methyl groups of the amine and hydrogens at C4 of the cyclohexene rings. This interaction is not present in 1a, 2a, and 3a, however, minimal interactions are possible between the amine methyl groups and H1.

Further studies incorporating more polar solvents are needed to elucidate the contribution of conformational relative to electronic effects on the NLO response of donor-π-acceptor chromophores.

EXPERIMENT

The following compounds were prepared according to literature procedures: 3-(dimethylamino)-2-cyclohexene-1-one (1b),[14] 5-dimethylamino-penta-2,4-dienal (2a),[15] 7-dimethylamino-hepta-2,4,6-trienal (3a).[15-17] Preparations for 1a, 2b, and 3b will be described in a future publication. As a means of calibrating the EFISH apparatus, the molecular hyperpolarizability was determined for 4-nitroaniline ($\beta_\mu = 7.1$ x 10^{-30} cm^5 esu^{-1} in CHCl$_3$). This compares well to previously reported values.[18] Molecular hypolarizabilities were determined by a concentration dependent analysis of the EFISH data. We performed all EFISH experiments at 1907 nm in chloroform.

REFERENCES

[1] G. A. Lindsay and K. D. Singer, *Polymers for Second-Order Nonlinear Optics* (ACS, Washington, DC, 1995).

[2] R. Dagani, Chemical & Engineering News **74** (10), 22-27 (1996).

[3] S. R. Marder, B. Kippelen, A. K. Y. Jen, and N. Peyghambarian, Nature **388** (6645), 845-851 (1997).

[4] T. Verbiest, S. Houbrechts, M. Kauranen, K. Clays, and A. Persoons, J. Mater. Chem. **7** (11), 2175-2189 (1997).

[5] S. R. Marder, L. T. Cheng, B. G. Tiemann, A. C. Friedli, M. Blanchard-Desce, J. W. Perry, and J. Skindhoj, Science **263** (5146), 511-514 (1994).

[6] I. D. L. Albert, T. J. Marks, and M. A. Ratner, J. Phys. Chem. **100** (23), 9714-9725 (1996).

[7] F. Meyers, S. R. Marder, B. M. Pierce, and J. L. Bredas, Chem. Phys. Lett. **228** (1-3), 171-176 (1994).

[8] S. R. Marder, C. B. Gorman, F. Meyers, J. W. Perry, G. Bourhill, J. L. Bredas, and B. M. Pierce, Science **265** (5172), 632-635 (1994).

[9] R. R. Pappalardo, E. S. Marcos, M. F. Ruizlopez, D. Rinaldi, and J. L. Rivail, J. Am. Chem. Soc. **115** (9), 3722-3730 (1993).

[10] J. L. Chiara, A. Gomezsanchez, and J. Bellanato, J. Chem. Soc., Perkin Trans. 2 (5), 787-798 (1992).

[11] Frequency-dependent hyperpolarizabilities were computed from the AM1 geometries using the ZINDO sum-over-states program (1907 nm, summed over the lowest 45 excited states using single pair excitations from the highest 12 occupied orbitals into the lowest 12 unoccupied orbitals.) (ZINDO, version 96.0 / 4.0.0, Biosym/MSI, San Diego CA, 1996).

[12] MOPAC version 6.0, Quantum Chemistry Program Exchange (QCPE #455), Department of Chemistry, Indiana University, Bloomington, Indiana.

[13] S. R. Marder, C. B. Gorman, B. G. Tiemann, and L. T. Cheng, J. Am. Chem. Soc. **115** (7), 3006-3007 (1993).

[14] C. J. Kowalski and K. W. Fields, J. Org. Chem. **46**, 197-201 (1981).

[15] S. S. Malhotra and M. C. Whiting, J. Chem. Soc. , 3812-3822 (1960).

[16] H. E. Nikolajewski, S. Dahne, and B. Hirsch, Chem. Ber. **100**, 2616-2619 (1967).

[17] A. C. Friedli, E. Yang, and S. R. Marder, Tetrahedron **53** (8), 2717-2730 (1997).

[18] Previous EFISH values for 4-nitroaniline: $\beta_\mu = 9.2$ x 10^{-30} cm^5 esu^{-1} in acetone,[19] 10 x 10^{-30} cm^5 esu^{-1} in N-methyl-2-pyrrolidinone,[20] 9.6 x 10^{-30} cm^5 esu^{-1} in 1,4-dioxane.[21]

[19] L. T. Cheng, W. Tam, S. H. Stevenson, G. R. Meredith, G. Rikken, and S. R. Marder, J. Phys. Chem. **95** (26), 10631-10643 (1991).

[20] L. T. Cheng, W. Tam, S. R. Marder, A. E. Stiegman, G. Rikken, and C. W. Spangler, J. Phys. Chem. **95** (26), 10643-10652 (1991).

[21] C. C. Teng and A. F. Garito, Phys. Rev. B **28**, 6766-6773 (1983).

A STUDY OF GROUP VELOCITY DISPERSION OF SINGLE-MODE OPTICAL WAVEGUIDE UTILIZING SILOXANE POLYMER

Y. UENO, K. KANEKO and S. TANAHASHI
KYOCERA corporation, R&D Center KEIHANNA
3-5 Hikaridai Seika-cho Soraku-gun Kyoto, 619-0237, Japan
Phone: +81-774-95-2125
Fax: +81-774-95-2119

ABSTRACT

We have developed a single-mode optical waveguide utilizing siloxane polymer on ceramic substrate for opto-electronic multichip modules (O/E-MCM) to operate Tera bps. In this paper, we describe group velocity dispersion for the optical waveguide utilizing siloxane polymer on ceramic substrate. The group velocity dispersion becomes 0 fs/cm·nm at around 1310 nm and 1530 nm wavelength. There is abnormal dispersion at around 1530 nm. The results showed that the group velocity dispersion of the optical waveguide was significantly related to the absorption spectrum of siloxane polymer. The group velocity dispersion of the optical waveguide is about 0.4 fs/cm·nm at 1550 nm wavelength. When a 500 femto seconds optical pulse, which has 16 nm for spectrum width, pass through the 50 mm long optical waveguide, pulse expansion is estimated 20 fs/cm·nm. This expansion of spectrum is well to distinguish femto seconds optical signals. Thus, the optical waveguide utilizing siloxane polymer is capable for femto seconds optical pulses transmission.

INTRODUCTION

Coming into the 21st century, telecommunication network has significant demands for high speed and large transmission capacity in the trunk line to achieve a developed multimedia society. To fulfill these demands, the Femtosecond Technology Research Association (FESTA) has been organized since 1995 and sponsored by New Energy and Industrial Technology Development Organization (NEDO) in Japan. In FESTA, ultrafast opto-electronics technology is developed to realize telecommunication network which is operated by femto second optical pulses. A packaging technique of opto-electronic multichip modules (O/E-MCM) are also developed to obtain a Tera bps level transmission system by members of FESTA. The system needs optical circuit with the least propagation loss and the least group velocity dispersion to transmit ultra short optical pulses.

Moreover, optical waveguide has to be formed on ceramic substrate which has excellent electrical characteristics in high frequency[1]. However, silica based optical waveguide fabricated on ceramic substrate has high loss, due to surface roughness of the ceramic substrate. Generally, a silica based optical waveguide is formed by Frame Hydrolysis Deposition (FHD) method operated at high temperature around 1000-1200 °C. When the silica based optical waveguide is fabricated on substrate with electrical circuit, its high temperature process damages the electrical circuit.

In order to solve these problems, a low temperature process has been developed to fabricate an optical waveguide on ceramic substrate utilizing siloxane polymer[2]. The optical waveguide utilizing the siloxane polymer has a propagation loss of 0.14 dB/cm[3] and is low value ranked next to that of silica based optical waveguide. This paper describes the group velocity dispersion of the siloxane polymer optical waveguide. The optical waveguide has the group

Mat. Res. Soc. Symp. Proc. Vol. 561 © 1999 Materials Research Society

velocity dispersion of 0 fs/cm·nm at 1310 nm, and that of 0.4 fs/cm·nm at 1550 nm wavelength. The siloxane polymer optical waveguide has capability of femto seconds optical pulses transmission.

MATERIAL AND PROCESS

-Material property -

Typical good properties of the siloxane polymer are shown in Table 1. Coupling loss between the optical waveguide utilizing siloxane polymer and silica based optical fiber is expected to be the least value due to that refractive index of the siloxane polymer is 1.444 at 1.31 μm wavelength, which is similar to that of silica. Optical transparency is higher than 99% at 1.31 μm wavelength for 4 μm thick siloxane polymer film. Fig.1 shows an IR absorption spectrum chart. There is an optical window around 1.31 among overtones vibration bands of C-H. Therefore, this resin has good optical properties to transmit optical signals. The siloxane polymer is supplied as an oligomer solution with organic solvent, and able to form a film by spin coating and curing at 270 °C. This process with low temperature does not damage a fabricated electrical circuit on the substrate. Moreover, the film has high thermal stability up to 400 °C, the curing temperature of polyimide. The waveguide can be fabricated on or under an electrical circuit composed of Cu/PI.

Table 1   Material property of siloxane polymer

Refractive index	1.444 (λ=1.3μm)
Optical transparency	>99% (λ=1.3μm)
Shrinkage for cure process	<5%
Curing temperature	270 °C
Thermal stability	400 °C

The siloxane polymer can improve rough surface of a substrate. For example, when the siloxane polymer thickness is 9 μm, surface roughness of the substrate is reduced to about 5% of that of alumina ceramic substrate (from 0.35 to 0.017 μm). The good coverage capability for surface roughness is coming from the siloxane polymer has very small shrinkage of around 5%, after curing process.

-Control of refractive index-

Generally, in order to make optical waveguides, core of optical waveguide is set higher refractive index than that of cladding. Refractive index of the siloxane polymer can be changed and controlled from 1.444 to 1.558 in addition of titanium-tetra-n-butoxide into the siloxane

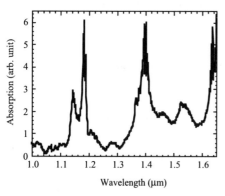

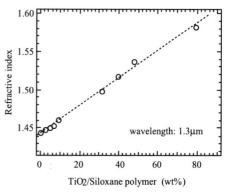

Fig.1   IR absorption spectrum chart of TiO$_2$ doped siloxane polymer.

Fig.2   Change in refractive index as a function of TiO$_2$ content in siloxane polymer.

oligomer before spin coating. Fig.2 shows a change of refractive index of siloxane polymer as a function of Ti contents. A desired refractive index can be easily obtained from its linear function.

-Manufacturing process of optical waveguide-

The manufacturing process flow of the optical waveguide is shown in Fig.3. First, the siloxane oligomer is coated with thickness of 9.0 μm for under cladding layer on an alumina ceramic substrate by spin coating and the TiO_2 doped siloxane oligomer coat of 7.0 μm thickness for core layer on the cladding layer. These layer are cured at 270 °C in N_2 gas atmosphere. Now, the thickness value of core and cladding layer mean that of the layers after cure process (Process 1). Aluminum for the metal mask pattern is fabricated on the core layer for the waveguide. Aluminum is deposited thickness of 0.5 μm by sputtering. Positive type resist is coated on the aluminum film, and patterned by photolithography (Process2). The aluminum film is patterned with reactive ion etching, and the resist removed (Process 3). The core layer is also formed by reactive ion etching (RIE) in O_2 and CF_4 gas. The mask material is chosen to have large selection rate between the mask and the siloxane polymer and found that aluminum has large selection rate of 14 (Process 4). After the siloxane polymer of 7.0 μm thick is etched, the aluminum mask is removed. And the siloxane polymer is coated by thickness of 5 μm for over cladding layer. (Process 5). The optical waveguide on the substrate is cut to expose waveguide's end to connect with optical fiber by dicing. An optimized dicing condition offers smooth surface of the facet as shown in Fig.4.

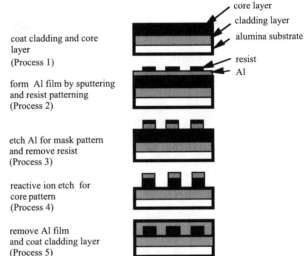

coat cladding and core layer
(Process 1)

form Al film by sputtering and resist patterning
(Process 2)

etch Al for mask pattern and remove resist
(Process 3)

reactive ion etch for core pattern
(Process 4)

remove Al film and coat cladding layer
(Process 5)

core layer
cladding layer
alumina substrate
resist
Al

Fig.3   Process flow of manufacturing of waveguide

MEASUREMENT OF GROUP VELOCITY DISPERSION

-Design of optical waveguide-

In order to make a measurement of group velocity dispersion on the siloxane polymer optical waveguide, straight type optical waveguide is formed on alumina ceramic substrate. The optical waveguide is designed to be 50 mm long, 7 μm wide and 7 μm high for core, 9 μm thick for under clad and 5 μm thick for upper clad. Refractive index are 1.4440 and 1.4482 for clad and core at 1.31 μm wavelength, respectively. (Δn=0.3%)

-Measurement method-

Group velocity dispersion of the optical waveguides is calculated from a group delay. Group delay is measured with Chromatic Dispersion Test System (HP 86037A). Measurement system is set up as block diagram in Fig.5. DUT (the siloxane polymer optical waveguide) is connected to 2000 mm long conventional single mode optical fiber with UV resin. Output pulses from the test system are passed through the optical fiber and DUT to input port. The output signals is modulated 2GHz in the test system. The test system indicates group delay from the different phase between the reference in the system and detected signals passed through the DUT. Group velocity dispersion can be obtained from the first differential of the group delay.

Fig.4 SEM image photograph of cutting section of the siloxane polymer optical waveguide on ceramic substrate with optimized dicing condition.

RESULT AND DISCUSSION

Fig. 6, 9 show measurement results of the group delay for wavelength at 1300 nm and 1550 nm wavelength band, respectively. The line indicates the curve fitted in with 3rd term Sellmeier's equation. The Sellmeier's equation is expressed as follows:

$$\text{Group Delay} = A\lambda^2 + B + C\lambda^{-2} \quad (1)$$

where, λ is wavelength and A,B and C are constants due to material property. At 1300 nm wavelength band, the group delay is almost on Sellmeier's curve, as shown in Fig. 6. Fig.7 shows the group velocity dispersion obtained from the first differential of the group delay. The group velocity dispersion becomes 0 fs/cm·nm at 1310 nm wavelength, and tendency to saturate. This feature is similar to that of silica based optical waveguide.

The group delay in entire measurement wavelength band, is roughly on the way to fit in with Sellmeier's curve. However, the group delay shifted upper and under of the Sellmeier's curve. This cause is abnormal dispersion due to the absorption. At around 1530 nm for wavelength, the curve of group delay greatly goes out of Sellmeier's curve and shows a concave shape. This shape of the curve around 1530 nm is similar to that of the output power for wavelength as shown in Fig. 8. The output power for wavelength indicates concave shape curve and minimum value at 1530 nm wavelength. This point means a peak of IR absorption spectrum at 1530 nm wavelength as shown in Fig.1.

The group velocity dispersion of the siloxane polymer optical waveguide has negative values between 1450 nm and 1530 nm wavelength, and becomes 0 fs/cm·nm at around 1530 nm wavelength as shown in Fig. 10, There is a small peak of IR absorption spectrum at 1550 nm wavelength. However the group velocity dispersion at 1550 nm wavelength takes maximum value of 0.4 fs/cm·nm. In 1 THz signals transmission system, 500 fs pulse is needed. 500 fs pulse of 1550 nm wavelength has 16 nm spectrum band width. Therefore, when the 500 fs pulse at 1550 nm wavelength is transmitted through the siloxane polymer optical waveguide (5 mm long), the pulse expanding is estimated 20 fs from integration of the dispersion between 1542 and 1558 nm as shown in Fig. 10, Therefore, because the expansion spectrum is small comparing with 500 fs's interval, the siloxane polymer optical waveguide the is applicable for Terabit O/E-MCM.

CONCLUSION

Estimation of the group velocity dispersion of the optical waveguide utilizing siloxane polymer fabricated on alumina ceramic substrate has been made. The group velocity dispersion becomes 0 fs/cm·nm at 1310 nm wavelength, and tendency to saturate. This feature is similar to

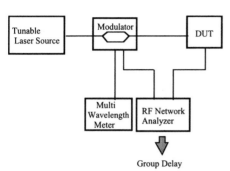

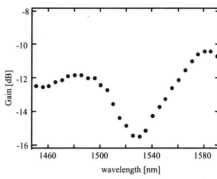

Fig.5 Chromatic Dispersion Test System block diagram

Fig.8 Output power for wavelength around 1500 nm

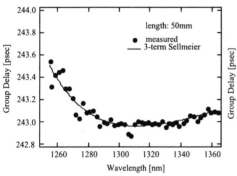

Fig.6 Group delay for wavelength around 1.300 μm

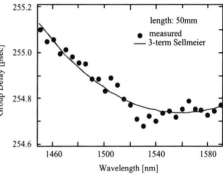

Fig.9 Group delay of the siloxane polymer optical waveguide for wavelength around 1500 nm

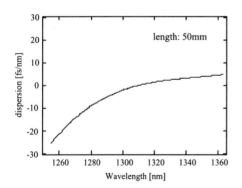

Fig.7 Group velocity dispersion of the siloxane polymer optical waveguide for wavelength around 1300 nm.

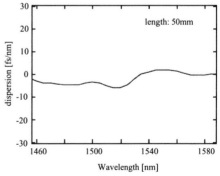

Fig.10 Group velocity dispersion of the siloxane polymer optical waveguide for wavelength around 1500 nm.

that of silica based optical waveguide. There is also 0 fs/cm·nm for the group velocity dispersion at 1534 nm waveguide. In 1550 nm wavelength band, the siloxane polymer optical waveguide the has abnormal dispersion, due to absorption feature of the material. At 1550 nm, the optical waveguide has maximum value for the group velocity dispersion. However, expansion of spectrum width due to this group velocity dispersion is small comparing with 500 fs's interval, the siloxane polymer optical waveguide the is applicable for Terabit O/E-MCM.

ACKNOWLEDGMENT

This development was performed under management of the Femtosecond Technology Research Association (FESTA) and supported by the New Energy and Industrial Technology Development Organization (NEDO).

REFERENCE

1. Y.Akahori, T. Ohyama, T. Hashimoto, and Y. Yamada, "Assembly and Wireing Technologies on PLC Platforms for Low-Cost and High-Speed Applications", Electronic Components and Technology Conference (ECTC), pp632-637, (1997)

2. Y. Ueno, K. Kaneko, and S. Tanahashi, " A New Single-mode Optical Waveguide on Ceramic Substrate utilizing Siloxane Polymer", Proc. SPIE **3289**, pp134-142, (1998)

3. S. Tanahashi, K. Kaneko, and M. Terasawa, "Single-Mode Optical Waveguide Using Siloxane Polymer on Cu-Polyimide Substrate", Electronic Components and Technology Conference (ECTC), pp189-193, (1995)

TWO PHOTON SPECTROSCOPY OF DITHIENYL POLYENES

THOMAS M. COOPER*, PAUL A. FLEITZ*, LAURA A. SOWARDS*, LALGUDI V. NATARAJAN**, SEAN KIRKPATRICK**, SURESH CHANDRA**, CHARLES W. SPANGLER†
*Air Force Research Laboratory, AFRL/MLPJ, 3005 P St. Ste. 1, Wright-Patterson Air Force Base, OH 45433
**Science Applications International Corporation, Dayton, OH 45434
†Department of Chemistry, Montana State University, Bozeman, MT 59717

ABSTRACT

To understand the properties of light-sensitive compounds used in optical limiters having photoinduced charge transfer mechanisms, we have investigated the photophysics of a series of di(2-thienyl-3,3'–butyl)polyenes. Spectroscopic measurements, including UV/Vis, fluorescence, fluorescence lifetimes, fluorescence quantum yields, triplet state lifetime, solvent effects and two-photon absorption coefficient were obtained as a function of the number of double bonds(n = 1-5). Trends in the data reflected the ordering, energy gap between and mixing of $^1B_u*$ and $^1A_g*$ excited state configurations.

INTRODUCTION

Optical limiters are devices containing dyes that undergo nonlinear absorption of light at high incident fluence. Most dyes investigated for optical limiting applications undergo significant population of the triplet state at high fluence. Examples include porphryins[1] and phthalocyanines[2]. An alternative optical limiting mechanism involves production of charge transfer excited states. The mechanism of optical limiting involves excitation of an electron donor in the presence of an electron acceptor forming a high optical density charge transfer state. Photogeneration of bipolaron-like dications has been proposed as a mechanism of charge-transfer based optical limiting. In particular, dithienyl polyenes have been proposed as donor chromophores to be used in these materials[3]. We have been investigating the photophysics of di(2-thienyl-3,3'-butyl)polyenes.

In this work we describe current spectroscopic measurements on a series of these polyenes, with n varying from 1 to 5. The compounds are designated as **I-V**.

EXPERIMENT

The dithienyl polyenes were synthesized by CWS according to published procedures[3]. UV/Vis spectra were measured with a Perkin-Elmer lambda 9 spectrophotometer. Fluoresence spectra were measured with a Perkin-Elmer LS-50B fluorescence spectrometer. Fluorescence lifetimes were measured with a Photon Technology International Laserstrobe fluorescence lifetime apparatus. Fluorescence quantum yields referenced to either quinine sulfate or perylene were measured according to published methods[4].

A single beam Z-scan measurement was also made at 800 nm using 100 fs pulses. Detectors D1 and D2 measure the open and closed aperture transmission through the sample respectively. Pump powers of 2-20 mW were used at 1 kHz. Reducing the repetition rate resulted in the same nonlinear coefficients to within the experimental error, therefore cumulative effects (such as thermal) can be ignored. The waist of the beam at focus was 130 μm, and the focal length of the lens was 1 m. It should be noted that a long focal length lens is used in order to allow higher pump powers, with larger beam waists, thereby preventing self phase modulation and subsequent continuum generation at the focus, while increasing the transmittance contrast. Sample path lengths were 5 mm and 2 mm, well with in the Raleigh range of the beam. Longer sample lengths are prohibited by self focusing of the beam within the sample, increasing the intensity above the threshold of white light generation. Two photon absorption cross sections are calculated from two photon absorption coefficients from the expression

$$\sigma_{TPA} = \frac{\beta h\nu}{N}$$

where N is the concentration in molecules/cm³, hv is the photon energy and β is the two photon absorption coefficient(cm/GW).

The triplet state of these compounds was probed by laser flash photolysis at 355 nm using the third harmonic of a Nd-YAG laser having a 7 nsec pulse width. The samples were deoxygenated by bubbling with N_2 during the experiment.

RESULTS AND DISCUSSION

The results we obtained can be interpreted from understanding of the electronic structure of polyenes[5] and diphenyl polyenes[6]. Figure 1 gives an energy level diagram for polyenes.

$$S_2 \underline{\qquad\qquad} \quad c_1\left|{}^1A_g^*\right\rangle - c_2\left|{}^1B_u^*\right\rangle$$

$$\Big\updownarrow \Delta E$$

$$S_1 \underline{\qquad\qquad} \quad c_1\left|{}^1A_g^*\right\rangle + c_2\left|{}^1B_u^*\right\rangle$$

$$S_0 \underline{\qquad\qquad} \quad \left|{}^1A_g\right\rangle$$

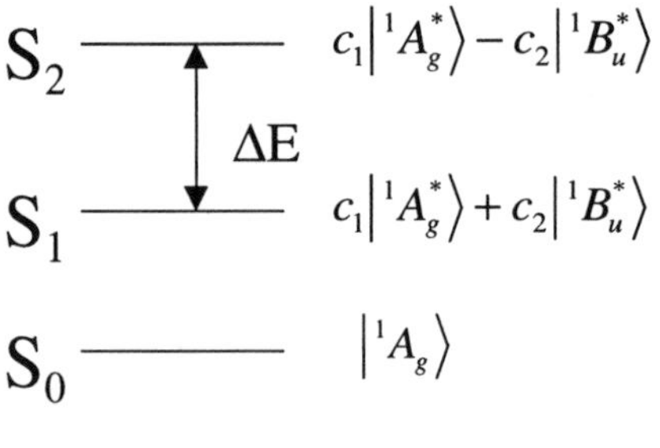

Figure 1. Energy level diagram for polyenes

All *trans*-polyenes have C_{2h} symmetry. The ground state has ${}^1A_g$ symmetry and two singlet excited states S_1 and S_2 have ${}^1B_u$ and ${}^1A_g$ symmetry. The one photon transition ${}^1A_g \rightarrow {}^1B_u^*$ is allowed while the transition ${}^1A_g \rightarrow {}^1A_g^*$ is forbidden. The ${}^1A_g^*$ state has high electron correlation, giving it "covalent" character while the ${}^1B_u^*$ state has lower electron correlaton, giving it more "ionic" character. Vibronic coupling with a low frequency b_u in plane bending mode results in the mixing of $^1A_g^*$ and $^1B_u^*$ configurations, making possible transitions including both the S_1 and S_2 states. The magnitude of the vibronic coupling increases as the energy (ΔE) between S_1 and S_2 decreases. When n = 1, the S_1 state has $^1B_u^*$ character($c_2 = 1$) and the S_2 state has $^1A_g^*$ character($c_1 = 1$). When n = 2, ΔE is small and S_1 and S_2 are nearly degenerate and there is maximum mixing between the $^1A_g^*$ and $^1B_u^*$ configurations. When n=3, S_1 has predominantly $^1A_g^*$ character($c_1 \approx 1$) and S_2 has $^1B_u^*$ character($c_2 \approx 1$). For n > 3, ΔE continues to increase, with S_1 having nearly 100% $^1A_g^*$ character($c_1 = 1$) and S_2 having nearly 100% $^1B_u^*$ character($c_2 = 1$). The state having predominantly $^1A_g^*$ character can be accessed by two photon excitation or through internal conversion from a nearby state having predominantly $^1B_u^*$ character. Upon photoexcitation, the various processes that can occur include internal conversion(IC), fluorescence(F), intersystem crossing(ISC) and photoisomerization from the *trans-trans* isomer to the *trans-cis* isomer(iso). The sum of the quantum yields for these various processes is approximately unity[6, 7].

$$\phi_{IC} + \phi_F + \phi_{ISC} + 2\phi_{iso} \approx 1$$

We applied these ideas in the interpretation of our dithienyl polyene data. Table I lists spectroscopic data for the series of dithienyl polyenes we investigated. As the number of double bonds increased, the absorption and emission maxima red-shifted, reflecting the increase in conjugation. The Stokes shift was estimated from the difference in excitation and emission maxima($\Delta E = \nu_{max}^{UV/Vis} - \nu_{max}^{Fl}$) and plotted in Figure 2. The increase in Stokes shift with n >2 suggested different absorbing and emitting states.

Table 1. Summary of Spectroscopic Data to Date

n^a	$\lambda_{max}^{UV/Vis}$ [b]	ε_{max} [c]	λ_{max}^{Fl} [d]	ϕ^e	τ_{Fl}^f	$\tau_{Triplet}^g$
1	358	25,069	424	0.023	1.27	0.25
2	378	42,216	451	0.030	0.67	3.1
3	398	65,842	512	0.063	5.79	2.5
4	416	96,985	561	0.0033	2.40	3.0
5	433	110,060	610	0.00039	0.41	2.9

[a]Number of double bonds
[b]Absorption maximum(nm) from UV/Vis spectrum of the dye dissolved in dichloromethane
[c]Extinction coefficient($\pm 10\%$) ($M^{-1}cm^{-1}$)
[d]Emission maximum(nm) from fluorescence spectrum
[e]Fluorescence quantum yield($\pm$ 15%)
[f]Fluorescence lifetime(nsec)
[g]Triplet state lifetime(μsec)

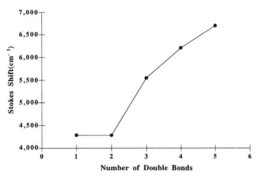
Figure 2. Stokes shift as a function of the number of double bonds.

We compared the UV/Vis and fluorescence spectra in a low polarizability solvent(hexane or octane) *vs.* benzene, a high polarizability solvent(Table 2). For compound **I**, the solvent-induced red shift was identical for both the UV/Vis and fluorescence spectra, suggesting an identical absorbing and emitting state. The red shift suggested polar $^1B_u^*$ character. In compound **II**, the red shift was somewhat smaller than that observed from the UV/Vis spectrum, implying the absorbing state had $^1B_u^*$ character while the emitting state had mixed $^1A_g^*$ and $^1B_u^*$ character. In compound **III**, there was a red shift in the UV/Vis spectrum but a negligible red shift in the fluorescence spectrum. This implied the absorbing state had $^1B_u^*$ character but the emitting state had predominantly $^1A_g^*$ character. From these results, emission occurred from the $^1B_u^*$ state in **I** while for **II** - **V**, there was $^1B_u^* \rightarrow {}^1A_g^*$ IC, followed by emission from the $^1A_g^*$ state.

Table 2. Solvent-Induced Shifts in Absorption and Fluorescence Spectra

n^a	$\Delta v(UV/Vis)^b$	$\Delta v(Fl)^c$
1	-312	-309
2	-424	-150
3	-445	-39

[a]Number of double bonds. For n = 1, 2 the solvent shift is from hexane to benzene. For n=3, the solvent shift is from octane to benzene.
[b]Shift in absorption maximum (cm^{-1})
[c]Shift in emission maximum (cm^{-1})

From the flash photolysis experiments we observed the triplet state in compounds **I-V**. Bubbling oxygen into the samples quenched the signal, giving evidence for triplet state formation. The highest signal-to-noise ratio was observed in compound **III**, suggesting compound **III** had the highest triplet state formation efficiency. For a related series of dithienyl polyenes, the maximum triplet formation efficiency was found for n=3[7]. Absorption spectra of the samples collected after the flash photolysis experiment showed evidence for photoisomerization in all five compounds.

The fluorescence quantum yield trend can be explained from published trends for similar dithienylpolyenes[7]. Photoisomerization was the dominant process for **I** and **II**. For compounds **III-V**, IC to S_0 predominated. We calculated the radiative rate constant,

$$k_F = \frac{\phi}{\tau_{Fl}}$$

and plotted the results in Figure 3. The rate constant for **I** was similar to that of stilbene[6], suggesting an allowed transition. The rate constant decreased through the polyene series, suggesting decreasing vibronic coupling between $^1A_g{}^*$ and $^1B_u{}^*$ configurations, leading to a lower oscillator strength for emission from the S_1 state.

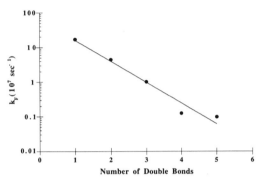

Figure 3. Radiative rate constant as a function of the number of double bonds.

The Z-scan experiments were performed with 100 fsec pulses. An example of Z-scan data for compound **II** is shown in Figure 4. Earlier experiments on these compounds involved 24 psec pulses[8]. We previously published two photon fluorescence data from dithienyl polyenes[9]. Nonlinear processes at this time scale are purely electronic, as IC to heat occurs on a nsec timescale[10]. Nonlinear absorption with nsec pulses includes thermal and photochemical effects. Table 3 lists two photon absorption coefficients measured to date. The data set is limited, but suggests no simple relation between the number of double bonds and the two photon absorption coefficient. No signal was observed for **I** or **III**. In **I**, the $^1A_g{}^*$ state is above the $^1B_u{}^*$ state, so excitation at 800 nm would yield a negligible two-photon absorption coefficient. Factors influencing the magnitude of the two photon absorption coefficient include the energy of the $^1A_g{}^*$ state and the bandwidth. We measured the two photon absorption coefficient at 800 nm. Excitation into the tail of the two photon absorption band will give a low absorption coefficient. Determination of the complete two photon absorption spectrum will clarify the observed trends.

Table 3. Two Photon Absorption Coefficients Measured to Date

Number of Double Bonds	σ_{TPA} (x10^{-50} cm^4 sec photon^{-1} molecule^{-1})
2	12,300
4	3,460
5	9,410

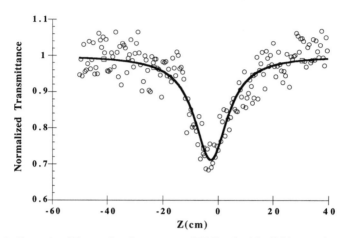

Figure 4. Example of Z-scan data for compound **II** dissolved in dichloromethane

CONCLUSIONS

The excited-state behavior of several all-*trans*-α,ω-dithienylpolyenes has been described in the literature[7-9]. Our data support the published observations and show that the electronic structure of dithienyl polyenes is similar to that of other polyenes. The two photon absorption coefficient of these compounds is fairly large and suggests synthesis of D-A-D structures from these compounds would give enhanced two photon absorption. Future work will involve investigation of the interaction of the spectroscopic properties of these compounds in the presence of charge transfer acceptors.

REFERENCES

1. W. Su, T.M. Cooper and M. C. Brant, Chem. Matl. **10**, 1212(1998).
2. J.W. Perry, K. Mansour, I.-Y. S. Lee, X.-L. Wu, P.V. Bedworth, C.-T. Chen, D. Ng, S.R. Marder, P. Miles, T. Wada, M. Tian and H. Sasabe, Science **273**, 1533(1996).
3. C.W. Spangler and M. He. Mat. Res. Soc. Symp. **479**, 59(1997).
4. J.N. Demas and G.A. Crosby, J. Phys. Chem. **75**, 991(1972).
5. G. Orlandi, F. Zerbetto and M.Z. Zgierski, Chem. Rev. **91**, 867(1991).
6. M.T. Allen and D.G. Whitten, Chem. Rev. ,**89**, 1691(1989).
7. G. Bartocci, A. Spalletti, R.S. Becker, F. Elisei, S. Floridi and U. Mazzucato, J. Am.Chem. Soc. , **121**, 1065(1999).
8. L.V. Natarajan, L.A. Sowards, C.W. Spangler, N. Tang, P.A. Fleitz, R.L. Sutherland and T.M. Cooper, Mat. Res. Soc. Symp., **479**, 135(1997).
9. L.V. Natarajan, S.M. Kirkpatrick, R.L. Sutherland, L.A. Sowards, C.W. Spangler, P.A. Fleitz and T.M. Cooper, SPIE Proc., **3472**, 151(1998).
10. N.J. Turro, *Modern Molecular Photochemistry*, (University Science Books, Sausalito,1991), p. 183.

COMBINED MAIN- AND SIDE-CHAIN AZOBENZENE POLYESTERS: A POTENTIAL FOR PHOTOINDUCED NONLINEAR WAVEGUIDES

F. SAHLÉN*, T. GEISLER*#, S. HVILSTED*, N. C. R. HOLME**, P. S. RAMANUJAM**, J. C. PETERSEN*

*Condensed Matter Physics and Chemistry Department, Risoe National Laboratory, DK-4000 Roskilde, Denmark
**Optics and Fluid Dynamics Department, Risoe National Laboratory, DK-4000 Roskilde, Denmark
*Danish Institute of Fundamental Metrology, DK-2800 Lyngby, Denmark

ABSTRACT

New combined main- and side-chain azobenzene polyesters, which exhibit an intensity dependent refractive index, have been prepared in order to optically fabricate nonlinear waveguides. Novel sulfone azobenzenes, a diester, (4-[[5-(ethoxycarbonyl)pentyl]sulfonyl]-4'-[[5-(ethoxycarbonyl)pentyl]methylamino] azobenzene, and a diol, 4-[[(8-hydroxy-7-methylhydroxy)-octyl]sulfonyl]-4'-N, N-dimethylamino azobenzene, have been used to prepare new polyesters by transesterification in the molten state. The polyesters have been characterized by UV-visible spectroscopy, differential scanning calorimetry (DSC), size exclusion chromatography (SEC), third harmonic generation (THG) and optical anisotropy measurements. The molar masses of the polyesters were in the range of 5000-10000 g mol^{-1}, which was sufficient in order to spin coat thin films. From THG measurements the polymers are shown to possess an off-resonance electronic $\chi^{(3)}$ of the order 10^{-12} esu corresponding to a nonlinear refractive index of 2.3×10^{-14} cm^2/W.

INTRODUCTION

Nonlinear waveguiding structures have aroused great interest due to their potential applications in the fields of optical switching, signal processing and communication[1,2,3]. Nonlinear waveguiding has also been used to study the interaction between biological molecules and the detection of specific chemical, biological and biochemical species[4]. Devices based on nonlinear waveguiding are also interesting for generation and transmission of nonclassical light[5].

The basic philosophy behind the current work is the fabrication of a NLO polymer in which nonlinear waveguides can be fabricated optically. The general design idea of our polymers comprises a $\chi^{(3)}$-active chromophore in the main chain, and a chromophore, which can be oriented by polarized light, in the side chain. The photoinduced orientation of the side chains results in anisotropy or birefringence, which provides the basis for the waveguide properties. Azobenzene containing polymers have shown that large birefringence can be induced in thin films of the polymers through polarized optical irradiation[6]. Refractive index differences between the ordinary and extraordinary waves of up to 0.1 have been reported in the literature[7]. It is also possible to provide in- and out-coupling to and from the waveguides through optically fabricated grating couplers. Large surface-relief gratings have been found to arise in azobenzene polymers, when they are irradiated with two orthogonally polarized beams[8,9,10]. This grating formation can be

---

Present address; Danish Defence Research Establishment, DK-2100 Copenhagen, Denmark

used to couple light into waveguides[11]. Thus for the first time, it is possible to fabricate a waveguide with input-output coupling entirely through optical means. We have carried out these ideas further and have fabricated a nonlinear optical polymer in which waveguides and in-and-output coupling can be created optically.

Even though the azobenzene by itself exhibits a third order nonlinearity, we have also incorporated a sulfone derivative in order to increase the overall nonlinearity. Para-toluene sulfonate (PTS) and bis-toluene sulfonate have been found to have very large third-order nonlinearities. PTS shows a strong cubic response with an observable quintic nonlinearity[12]. It has been shown that the large cubic nonlinearity is capable of producing index changes approaching 10^{-2} for incident intensities of a few gigawatts/cm^2. Fabrication of channel waveguides in para-toluene sulfonate has been reported[13]. The potential of PTS for optical switching devices has been examined by Lawrence et al.[14]. Based on these facts, it was surmised that introduction of sulfone as a substituent on the azobenzene will give rise to a large third-order nonlinearity. Furthermore, making this molecule as a part of a main chain, with substituted or unsubstituted azobenzene in the side chain in a polymer system was expected to produce a system in which it will be possible to fabricate waveguiding structures optically. The concentration of the nonlinear chromophore can be high. Phase separation effects and difficulties in having to prepare single crystals can thus be avoided. The polymer can be spin-coated making fabrication of high quality of optical films.

Like π-electron-conjugated polymers, both chromophore-functionalized side-chain and main-chain polymers seem promising for third-order nonlinear optics[15]. The chromophore concentration can be increased by attaching the active molecules to the polymer chain compared to doped polymers (guest-host systems) where the concentration is limited by aggregation problems. It has previously been demonstrated that chromophore functionalized polymers can posses appreciable third-order susceptibilities approaching those of fully conjugated backbone polymers. Both dye attached side-chain[16,17,18] and dye-introduced main-chain[17] polymers have been studied. The azo dye attached copolymers have larger $\chi^{(3)}$ values than the stilbene dye attached copolymer[17]. These studies demonstrate that high $\chi^{(3)}$ values can be obtained in relatively short π-electron-conjugated chromophore-functionalized polymers by introducing a dye molecule into a polymer backbone and that processable polymeric materials with $\chi^{(3)}$ values higher than 10^{-10} esu can be obtained.

EXPERIMENT

Instrumentation

UV-VIS spectra were measured on a Varian Cary 1E UV-visible spectrophotometer. The molar masses of the polyester samples of 2.5 mg/mL concentration were determined by size exclusion chromatography (SEC), using a Viscotek Model 200 detector. Tetrahydrofuran was used as eluent and the separation was made on two columns of Polymer Laboratories gel mixed-D. Differential scanning calometry (DSC) data were obtained on a Perkin-Elmer Pyris 1 with a scanning rate of 10 °C/min. Typically 4 mg of polymers were used for glass transition temperature detection. The transition temperatures were determined from the second heating/cooling traces. An Olympus BH-2 equipped with a Mettler FP82 Hot Stage and a Mettler FP90 Central Processor was used for polarizing optical microscopy (POM) in order to reveal any liquid crystalline phases.

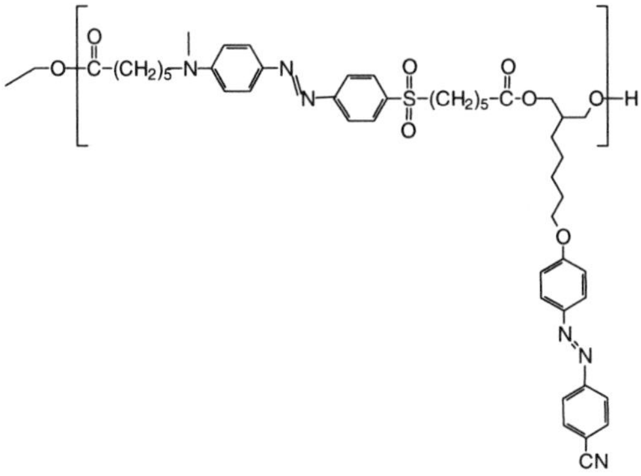

Figure 1. The chemical structure of **P5-13**.

Synthesis

The new polymers were designed with a $\chi^{(3)}$ active chromophore in the main-chain and an azobenzene chromophore in the side-chain, which could be oriented by polarized light (Fig. 1). Three polyesters (**P5-12**, **P5-13** and **P5-14**) were prepared by transesterfication in the molten state of a diester monomer (4-[[5-(ethoxycarbonyl)pentyl]sulfonyl]-4'-[[5-(ethoxycarbonyl)-pentyl]methylamino] azobenzene (**5**)) and three different diol monomers (4-[[(8-(hydroxy-7-methylhydroxy)octyl]sulfonyl]-4'-*N*, *N*-dimethylamino azobenzene (**12**), [2-[6-[4-(4'-cyano-phenyl)azo]phenoxy]hexyl]-1, 3 propanediol (**13**) and [3-[4-(4'-cyanophenyl)azo]phenoxy]-1, 2 propanediol (**14**)). The synthesis and characterization of the new sulfoneazobenzene monomers, **5** and **12**, will be presented elsewhere. A fourth polyester **P15-12** was similarly prepared by transesterification of diphenyl tetradecanedioate and **12**. In figure 1 the chemical structure of **P5-13** is shown. **P5-13** consists of two different azobenzene chromophores placed in the main chain and in the side chain, and a six-methylene-group side-chain spacer. The chromophore in the main chain has a sulfone group as electronic accepting part and an amine as electron donating part. The chemical structure of polyester **P5-12** comprises of the same chromophore in the main chain as for **P5-13** but in the side chain a chromophore identically to the one in the main-chain was incorporated. Polyester **P5-14** differs from the other polyesters in the diol part that comprises an ethylene glycol with only one methylene group spacing the azobenzene moiety. **P5-12**, **P5-13** and **P5-14** are amorphous polymers, exhibiting glass transition temperatures at 50°C, 42°C and 52°C respectively. The polymer **P5-13** shows two absorption maxima (λ_{max}) at 367 nm and 452 nm, corresponding to the absorption of the respective side-chain and the main-chain chromophore. The cut-off absorption (λ_{cutoff}) is about 540 nm. The absorption spectrum of **P5-14** was almost identical to the one of **P5-13**, showing λ_{max} at 363 nm and 453 nm. **P5-12** exhibited one absorption maximum at 450 nm and a λ_{cutoff} at 540 nm. The molar masses of the polyesters ranged from 5.000-10.000 g mol^{-1}, which was sufficient for spin coating thin films. Films were spun onto glass substrates from chloroform solutions with a concentration of 20 mg/200 µL to 20 mg/800 µL, resulting in a film thickness of 1.50 µm down to 0.10 µm. The thickness measurements were made using a stylus profilometer.

THG Measurements

The THG Maker fringe technique was applied in order to determine the purely electronic contribution to the third-order nonlinear optical susceptibilities ($\chi^{(3)}$) of these polymer samples. Off-resonance conditions were assured by the use of a fundamental wavelength of 1907 nm obtained by Raman shifting the 1064 nm pulses from a Q-switched Nd:YAG laser in a 1-m high pressure (27 bar) hydrogen cell. The vertically polarized beam was weakly focussed on the sample, which was mounted on a rotation stage in a vacuum chamber to eliminate the THG contribution from air. The third-harmonic beam (636 nm) was filtered from other wavelengths by a combination of IR-filters and a monochromator and detected in a transmission geometry using a photomultiplier in connection with a photon counter. Maker fringes were recorded in the range from -50° to 50° with respect to normal incidence, with the film facing the detector. First a fringe pattern of a sample consisting of a thin film on substrate was recorded. Both the film and substrate contribute to the third-harmonic signal. After removal of the thin film, *in situ*, the contribution from the substrate was determined by recording a fringe pattern under the same experimental conditions.

RESULTS

The $\chi^{(3)}$ values of the thin polymer films were calculated from the following equation

$$\chi_f^{(3)} = \frac{2}{\pi} \frac{l_{c,s}}{l_f} \left(\frac{I_{3\omega,f}}{I_{3\omega,s}} \right)^{1/2} \chi_s^{(3)} \qquad (1)$$

where it is assumed that the film thickness, l_f, is much thinner than the THG coherence length, $\chi_s^{(3)}$ and $l_{c,s}$ are the third-order nonlinear susceptibility and the THG coherence length of the glass substrate assumed to be 4.4×10^{-14} esu[19] and 18 µm, respectively. $I_{3\omega,s}$ is the third-harmonic intensity from the substrate obtained from the envelope function at normal incidence, when no film is present. $I_{3\omega,f}$ is the third-harmonic intensity from the thin film sample, corrected for the presence of the substrate[20]. The results are presented in Table I. The number density, N, of repeat units is calculated assuming a polymer density of 1.18 g/cm³.

Table I

Polymer	λ_{max} (nm)	$\chi^{(3)}$ (10^{-12} esu)	N (10^{20} cm⁻³)	$\langle\gamma\rangle$ (10^{-36} esu)
P5-13	367,452	1.2	8.4	360
P5-12	450	1.3	7.7	420
P5-14	363, 453	1.7	9.3	460
P15-12	450	0.69	10.6	160

The $\chi^{(3)}$ values are obtained from films with thicknesses in the range 0.12-0.16 µm except for the last, where it was 0.55 µm in order to obtain a significant signal. These films can safely be considered much thinner than the coherence length. The nonlinearity per repeat unit, $\langle\gamma\rangle$, is calculated from $\chi^{(3)} = LN\langle\gamma\rangle$, where L is a correction factor due to local-field effects. We assume a value of L=4 obtained from the Lorentz approximation with refractive indices close to 1.5.

The three polymers with chromophores in both the main and side chain possess $\chi^{(3)}$ values between 1.2 and 1.7×10^{-12} esu. These values are not considered to be significantly different considering the experimental uncertainty from e.g. the thickness measurements. Films with thickness just below 0.5 µm showed approximately 10% lower nonlinearity. It must be remarked that the observed nonlinearity arises both from the azobenzene chromophores and the sulfone group. Further analysis of the fringe pattern show that the phase of the nonlinearity is close to zero within 25 degrees such that the real part is positive. In the analysis the polymer refractive indices are assumed to posses normal dispersion such that $n_{3\omega}$ is larger than n_ω.

In the following these values are compared with those obtained in the literature for azobenzenes and toluene sulfonate. The third-order nonlinear optical response of a thin film of Disperse Red 1 (DR1) functionalized PMMA has been studied by Rangel-Rojo et al. using the z-scan technique[21] with tuneable picosecond pulses. A nonlinear refractive index $n_2 = -5 \times 10^{-9}$ cm^2/W, corresponding to a Re$\chi^{(3)}$ = -3.0×10^{-15} m^2/V^2 (-2.1×10^{-7} esu), has been measured at 570 nm (number density 5.6×10^{20} cm^{-3}). This value decreases 500 times at 610 nm. The measured near-resonant nonlinearity, $|\chi^{(3)}| = 10^{-7}$ esu, is indeed very large, and is comparable with what has been observed near resonance for single crystal polydiacetylenes. The large value is believed to be associated with the photoinduced trans-cis isomerization of the azo dye. Sub-picosecond resonant third-order nonlinearity of 4-nitro, 4'-(N-methyl, N-hexadecyl) aminoazobenzene doped polymer film was investigated by Dong et al.[22] They obtained a $\chi^{(3)}$ of 6.3×10^{-12} esu and a dephasing time of 1×10^{-12} s respectively by means of transient degenerate four-wave mixing and z-scan techniques. These values compare favorably with our values obtained above. It must be emphasized that third harmonic generation measurement such as ours records only electronic contributions.

It is clear that the nonlinearity per repeat unit is larger for the polymers with two chromophores per repeat unit. The nonlinearity per chromophore is close to 200×10^{-36} esu. This is in agreement with the values reported by Lap-Tak Cheng et al.[23] for azobenzene chromophores like Disperse-Red (DR1) for which a $\langle \gamma \rangle = 190 \times 10^{-36}$ esu is reported.

However, the $\chi^{(3)}$ values are lower than reported for DR1 functionalized PMMA. A polymer with a dye content of 62 wt% was reported[23] to have a $\chi^{(3)} = 6.8 \times 10^{-12}$ esu at 1.9 µm. In this case, however, the number density is larger than in our case.

Lawrence et al.[12] have shown that one dimensional conjugated poly[2,4-hexadiyne-1,6-diol-bis-p-toluene-sulfonate] has a nonlinear response of $n_2 = 2 \times 10^{-12}$ cm^2/W at 1.6 µm. D. Y. Kim et al.[24] have measured the nonlinear Kerr coefficient $n_2 = 3.2 \times 10^{-12}$ cm^2/W in a single crystal of p-toluene sulfonate. Our observed value for $\chi^{(3)}$ of the order of 10^{-12} esu, corresponding to a nonlinear refractive index of 2.3×10^{-14} cm^2/W is approximately two orders of magnitude smaller. One of the reasons is certainly the packing of the sulfone azobenzene chromophores in our polymers is not optimum. Secondly, while z-scan measurements give a value for a $(\omega;\omega,-\omega,\omega)$ process, third harmonic generation gives the value for a $(3\omega;\omega,\omega,\omega)$ process.

For conjugated poly(1,6-heptadiyne) having active NLO chromophores of e.g. 4-(dimethylamino)-4'-[[3-(didropargylacetoxy)propyl]-sulfonyl]stilbene as side-chains the $\chi^{(3)}$ measured by THG at 1907 nm were found to be 2.6×10^{-11} esu[26]. However, an even larger value was obtained for the PTDPA backbone alone. The nonlinearity of the backbone was little affected by the pendent NLO chromophores.

CONCLUSIONS

A polyester containing a sulfone azobenzene in the main and or side chain has been fabricated. The measured third order nonlinearities are comparable to those obtained off resonance for other chromophore functionalized polymers. Initial measurements of the optical anisotropy show that it is possible to create a large birefringence, paving the way for fabrication of waveguides. Surface relief gratings to facilitate in- and out-coupling of light into waveguides have also been made in the polyester. It is hoped that in the near future an all optical fabrication of nonlinear waveguides in the polyester will be feasible.

REFERENCES

1. A. Barthelmy, S. Maneuf and C. Froehly, Optics Commun. **55**, 193 (1985).
2. J.S. Aitchison, A.M. Weiner, Y. Silberberg, M.K. Oliver, J.L. Jackel, D.E. Leaird, E.M. Vogel and P.W. Smith, Opt. Lett. **15**, 471 (1990).
3. G.R. Allan, S.R. Skinner, D.R. Andersen and A.L. Smirl, Opt. Lett. **16**, 156 (1991).
4. B.J. Luff, R.D. Harris, J.S. Wilkinson, R. Wilson and D.J. Schiffrin, Opt. Lett. **21**, 618 (1996).
5. N. Korolkova and J. Perina, Optics Commun. **136**, 135 (1996).
6. M. Eich, J.H. Wendorff, B. Rech and H. Ringsdorf, Macromol. Chem. Rapid Commun. **8**, 59 (1987).
7. S. Hvilsted, F. Andruzzi, C. Kulinna, H.W. Siesler and P.S. Ramanujam, Macromolecules, **28**, 2172 (1995).
8. P. Rochon, E. Batalla and A. Natansohn, Appl. Phys. Lett. **66**, 136 (1995).
9. D.Y. Kim, S.K. Tripathy, L. Li and J. Kumar, Appl. Phys. Lett. **66**, 1166 (1995).
10. P.S. Ramanujam, N.C.R. Holme and S. Hvilsted, Appl. Phys. Lett. **68**, 1329 (1996).
11. P. Rochon, A. Natansohn, C.L. Callender and L. Robitaille, Appl. Phys. Lett. **71**, 1008 (1997).
12. B.L. Lawrence, M. Cha, J.U. Kang, W. Torruellas, G. Stegeman, G. Baker, J. Meth and S. Etemad, Electron. Lett. **30**, 447 (1994).
13. D.M. Krol and M. Thakur, Appl. Phys. Lett. **56**, 1406 (1990).
14. B.L. Lawrence, M. Cha, W.E. Torruellas, G. Stegeman, S. Etemad, G. Baker and F. Kajzar, Appl. Phys. Lett. **64**, 2773 (1994).
15. H.S. Nalwa, in *Nonlinear Optics of Organic Molecules and Polymers*, edited by H.S. Nalwa and S. Miyata (CRC Press Inc, Boca Raton, 1997) p. 718.
16. S. Matsumoto, K.-I. Kubodera, T. Kurihara and T. Kaino, Appl. Phys. Lett. **51**, 1 (1987).
17. S. Matsumoto, T. Kurihara, K. Kubodera and T. Kaino, Mol. Cryst. Liq. Cryst. **182A**, 115 (1990).
18. W.R. Torruellas, R. Zanoni, M.B. Marques, G.I. Stegeman, G.R. Mohlmann, E.W.P. Erdhuisen and W.H.G. Horsthuis, Chem. Phys. Lett. **175**, 267 (1990); J. Chem. Phys. **94**, 6851 (1991).
19. D. Morichère, M. Dumont, Y. Levy, G. Gadret and F. Kajzar, SPIE **1560**, 214 (1991).
20. T. Geisler, K. Pedersen, A.E. Underhill, A.S. Dhindsa, D.R. Greve, T. Bjørnholm and J.C. Petersen, J. Phys. Chem. B **101**, 10625 (1997).
21. R. Rangle-Rojo, S. Yamada and H. Matsuda, Appl. Phys. Lett. **72**, 1021 (1998).
22. F. Dong, E. Kousoumas, S. Couris Y. Shen, L. Qiu and X. Fu, J. Appl. Phys. **81**, 7073 (1997).
23. L.-T. Cheng, W. Tam, S.H. Stevenson, G.R. Meredith, G. Rikken and S.R. Marder, J. Phys. Chem. **95**, 10631 (1991).
24. F. Kajzar and M. Zargoska, Nonlinear Opt. **6**, 181 (1993).
25. D. Y. Kim, B. L. Lawrence, W. E. Torruellas, G. I. Stegeman, G. Baker and J. Meth, Appl. Phys. Lett. **65**, 1742 (1994).
26. H.-J. Lee, S.-J. Kang, H.K. Kim, H.-N. Cho, J.T. Park and S.-K. Choi, Macromolecules, **28**, 4638 (1995).

DESIGN AND SYNTHESIS OF DENDRIMERS WITH ENHANCED TWO-PHOTON ABSORPTION

E. H. Elandaloussi*, C. W. Spangler*, C. Dirk**, M. Casstevens***, D. Kumar***, R. Burzynski***
*Department of Chemistry and Biochemistry, Montana State University, Bozeman, MT 59717
**Department of Chemistry, University of Texas-El Paso, El Paso, TX
***Laser Photonics Technology, Inc., 1576 Sweet Home Rd., Amherst, NY 14228

ABSTRACT

Model dendrimers based on bis-(diphenylamino)stilbene repeat units have been synthesized and shown to have very stable bipolaronic charge states. Calculations at the AM-1 level predict two orders magnitude enhancement in the third-order nonlinearity in going from the neutral to bipolaron state. Measurement of two-photon cross-sections for PPV dimers with attached diphenylamino donor groups show enhanced two-photon cross-sections for ns pulses. Preliminary evaluation of one of the model dendrimers indicates even larger two-photon cross-sections are possible.

INTRODUCTION

During the past few years dendrimers have displayed a number of unique properties, in particular their propensity to form globular structures in the higher generations. Such structures may have unique photonic properties since almost all photons impinging on the dendrimer structure may encounter an absorbing unit with correct conformation for efficient reaction. We have previously described[1] the formation of two model dendrimers based on bis-(diphenylamino)stilbene units, and showed that they formed stable bipolaronic dications that were considerably red-shifted from the bis-(diphenylamino)stilbene parent. We would now like to report on some preliminary evaluation of their photonic properties.

SYNTHESIS OF MODEL DENDRIMERS

Dendrimers can be formed by two essentially different routes. Divergent approaches rely on successive structural iterations on branched structures which may have several arms to allow synthetic attachment of new repeat units[2]. Thus, each succesive dendrimer generation increases rapidly in number of active units, surface functionalities, molecular weight and size (diameter). Convergent approaches build from the outside in, with pieces of the dendrimer structure (dendrons) eventually being attached to a multifunctional core unit to achieve the final dendrimer structure. In our work we have utilized a divergent strategy to produce two model dendrimers based on bis-(diphenylamino)stilbene repeat units. Model dendrimer **1** has three "arms" while model dendrimer **2** has four. The structure of model dendrimer **2** is illustrated below. The synthesis of the G-0 Model Dendrimer **1** is illustrated in Figure 1.

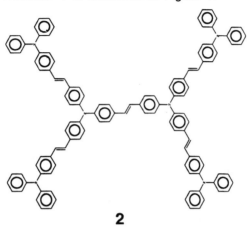

2

Oxidation of both dendrimers was carried out as we have described previously[3] with SbCl$_5$ in solution, obtaining stable bipolaronic dications whose spectra show large red shifts in oscillator strength, and are red-shifted when compared to the parent bis-(diphenylamino)stilbene parent structure [3]. These results are shown in Table 1.

Table 1. Absorption Characteristics of Neutral and Oxidized Dendrimer Models

Compound	π-π^* neutral (nm)	π-π^* oxidized (nm)
1	406	716
2	423	710
3	379	688

Results and Discussion

Dirk has recently proposed[4] that 2-d or 3-d electron delocalization of charges in organic structures should lead to large enhancement of the third-order nonlinear hyperpolarizabilities. This is particularly true for

Figure 1. Synthesis of G-0 Model Dendrimer

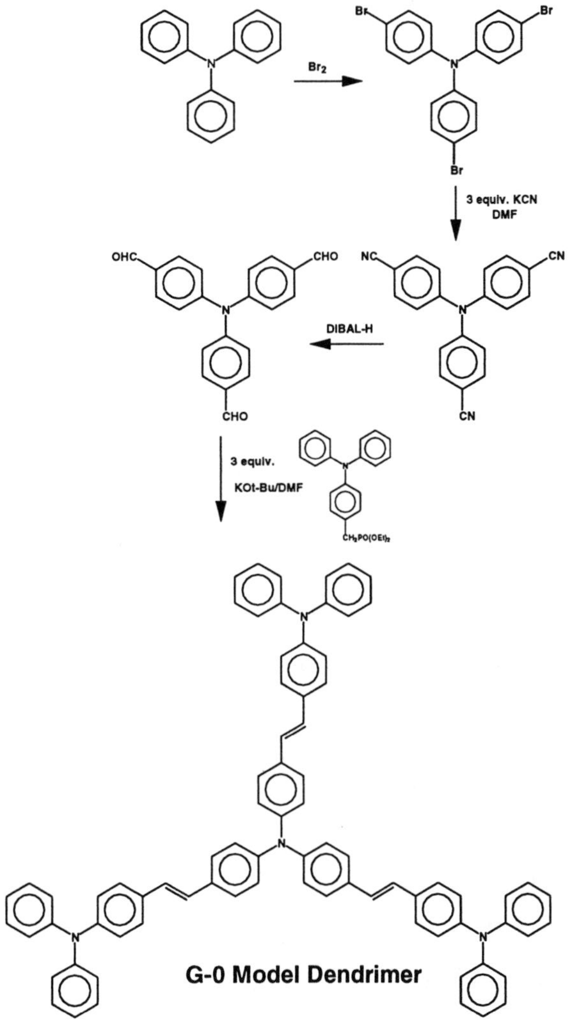

G-0 Model Dendrimer

the +2 bipolaron where γ turns rapidly negative and achieves huge values as the size of the delocalization regime increases and dwarfs the typical quasi-1-d experimental systems. These calculations, carried out at the AM-1 level, also predict that the ratio of γ to α also increased dramatically with increasing 3-d delocalization, which indicates that the commonly referred to nonlinearity-absorptivity trade-off may not exist in these systems. Calculations on model dendrimer **1** predicts a gigantic value for the bipolaron as shown below:

γ_0 (neutral) $= 1.51 \times 10^{-33}$ esu γ_0 (+2 BP) $= 4.67 \times 10^{-31}$ esu

The predicted two oders of magnitude enhancement may have important consequences for a varity of photonic applications, as we will discuss below for design criteria for new chromophores with enhanced two-photon absorption cross-sections.

We have recently designed several new chromophores based on PPV-dimer, which we believed would have enhanced two-photon cross-sections for use in optical power limiting of nanosecond (ns) pulses. Perry, Marder and coworkers have recently shown that **3** has an unusually large two-photon cross-section, and more recently have elucidated their design approach to new TPA chromophores[5,6]. We had simultaneously been involved in also designing several new TPA chromophores in a collaboration between Laser Photonics Technology and Montana State University, and two of the more promising candidates are shown below. Their effective two-photon cross-sections for ns pulses are quite large, and indicative of excited state absorption, driven by the TPA ,that has proven to be of use for optical limiting of the ns pulses. TPA-1 is an effective limiter of ns pulses, with its peak behavior at 670 nm.

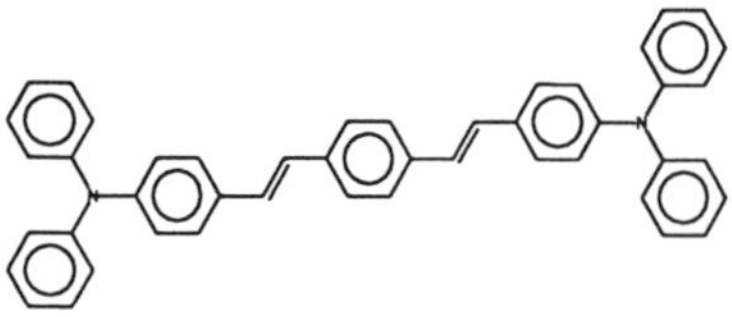

$\sigma_2 = 6{,}670$ (10^{-50} cm^{4}s/photon)

$\sigma_2 = 8{,}180$

What is particularly attractive about dendrimers such as **1** and **2** is the possibility of these materials being effective optical power limiters (OPLs). Recent design concepts for enhanced two-photon cross-sections have been proposed by Reinhardt and coworkers[7], which relates the two-photon cross-sections to the Im component of $\chi^{(3)}$:

$$\sigma_2 = \frac{8\pi^2 h\nu \; \text{Im} \; [\chi^{(3)}]}{n^2 c^2 N}$$

where h = Planck's constant, ν = frequency of incident light, n = index of refraction, c = speed of light in vacuum and N = number of absorbing molecules.

Photogeneration of charge states also leads to significant increases in photo-induced absorption (PIA)[8]. Two-photon absorption is an attractive entry into the excited state manifold wherein structural relaxation of the initially formed singlet S_1 state leads to highly absorbing polaronic or bipolaronic transient states. Such a scenario may provide extremely effective optical power limiting of ns pulses.

Preliminary measurement in solution of the effective two-photon cross-section of dendrimer **2** utilizing ns pulses has recently been carried out by He and Prasad at SUNY-Buffalo. The result is one of the largest effective cross-sections we have seen:

$$\sigma_2 = 1,000 \quad (\times \; 10^{-48} \; cm^4/photon)$$

Such a large value for σ_2 indicates significant excited state absorption. We are pursuing further studies to elucidate the underlying mechanism giving rise to this extraordinarily large TPA cross-section.

Conclusions

We have shown that dendrimer models based on bis-(diphenylamino)stilbene repeat units and various core configurations can be synthesized by divergent strategies. Oxidative doping of these materials to stable bipolaronic states suggest their photonic properties might be useful for applications involving transient nonlinear absorption (e.g. reverse saturable absorption). Dendrimers based on photo-active repeat units may also be extremely efficient two-photon absorbers. Preliminary two-photon cross-section measurements in collaboration with Prof. Paras Prasad's group indicate an extraordinarily large effective cross-section for dendrimer **2**, and will need be studied in more detail to confirm the magnitude of this cross-section. Other similar structures will also need to be explored to see if they, too, have such large cross-sections. Detailed measurements are also in progress to delineate the wavelength profile for the TPA, and the wavelength for maximum optical limiting. Work is also underway to incorporate TPA-1 and TPA-2 chromophores as dendrimer repeat units to determine if incorporation into a 2-d structure will further enhance their large TPA cross-sections.

Acknowledgements

Partial support of this research by the Air Force Office of Scientific Research under grants F49620-96-1-0440 and F49620-97-1-0413 is gratefully acknowledged.

References

1. C. W. Spangler and E. H. Elandaloussi, Polym. Preprints **39(2)**, p. 1055 (1998).

2. A. Tomalia and H. D. Hurst, in *Topics in Current Chemistry 165: Supramolecular Chemistry I – Directed Syntheses and Molecular Recognition*, edited by E. Weber, Springer-Verlag, Berlin, 1993, pp. 193-313.

3. C. W. Spangler in *Handbook of Conducting Polymers*, edited by T. A. Skotheim, R. L. Elsenbaumer and J. R. Reynolds, Marcel Dekker, Inc., New York, 1998, pp. 743 - 763.

4. O. Xie and C. W. Dirk, J. Phys. Chem. (in press).

5. J. E. Ehrlich, X. L. Wu, I.-Y. S. Lee, A. A. Heikal, Z.-Y. Hu, H. Rockel, S. R. Marder and J. W. Perry, Mat. Res. Soc. Symp. Proc. **479**, p. 9 (1997).

6. M. Albota, D. Beljonne, J.-L. Bredas, J. E. Ehrlich, J.-Y. Fu, A. A. Heikal, S. E. Hess, T. Kogej, M. D. Levin, S. R. Marder, D. McCord-Maughon, J. W. Perry, HJ. Rockel, M. Rumi, G. Subramanian, W. W. Webb, X.-L. Wu and C. Xu, Science **281**, p. 1653 (1998).

7. B. A. Reinhardt, L. L. Brott, S. J. Clarson, A. G. Dillard, J. C. Bhatt, R. Kannan, L. Yuan, G. S. He and P. N. Prasad, Chem. Mater. 10, p. 1863 (1998).

8. X. F. Cao, J. P. Jiang, R. W. Hellwarth, L. P. Yu, M. Chen and L. R. Dalton, Proc. SPIE **1337**, p. 114 (1990).

PHOTONIC APPLICATIONS OF BIPOLARON FORMATION IN BIS-(DIPHENYLAMINO)DIPHENYLPOLYENE

K. ASHWORTH*, B. REEVES*, A. FROST*, C. SPANGLER*
Department of Chemistry and Biochemistry, Montana State University, Bozeman, MT 59717

ABSTRACT

Bis-(dialkylamino) and bis-(diphenylamino) substituted diphenylpolyenes have recently been shown to have exceptionally large two-photon cross-sections, and have also been used in optical power limiting (OPL) applications. Another potential mechanism for OPL in these materials involves the photogeneration of highly absorbing polaronic radical-cations or bipolaronic dications which can then function as highly absorbing reverse saturable absorbers (RSAs). The synthesis of series of these unique polyenes has been carried out, as well as studies of chemically induced bipolaron formation in solution. In order to actually be used in OPL devices, it is probable that these chromophores will need to be incorporated into polymer formulations, In this paper we describe three distinct ways that these chromophores can be incorporated in polymer structures: (1) as pendant groups attached to a PMMA backbone, (2) as formal repeat units in a copolymer, and (3) perhaps the most interesting, as surface attached functionalities in dendrimer formulations. The formation of bipolaron charge species in these polymers will be addressed.

INTRODUCTION

During the past 10-15 years our research group has synthesized a large number of diphenyl and dithienyl polyenes which have been examined from the perspective of establishing structure-property relationships in order to design materials with increased optical nonlinearity[1-3]. One of the ways of increasing the third-order optical nonlinearity in these systems is by chemical oxidation, which introduces highly delocalized polaronic radical-cations or bipolaronic dications into the conjugation sequence[4]. These charge states can be stabilized by inclusion of mesomerically interactive electron-donating substituents on the aromatic ring termini, illustrated below:

Recently we have reported on the synthesis of a series of bis-(diphenylamino)diphenylpolyenes[5], which have rather unique properties. The substitution of phenyl rings for alkyl groups on the amino functionality greatly increases their solubility, and we have synthesized polyenes with up to 10 double bonds, which is impossible for simple alkyl groups due to their increasing insolubility. The phenyl substituents also increase the thermal and photochemical stability. Upon chemical oxidation these polyenes yield very stable bipolaronic dications. Previous to this study, we were not able to form stable bipolarons in and series with less than 4 conjugated double bonds in either the diphenyl or dithienylpoyenes. However, with diphenylamino stabilizing groups, even 1,2-bis-(diphenylamino)-E-stilbene forms a stable bipolaron. We have also synthesized diphenylpolyenes with one diphenylamino functionality and one N-ethyl-Nhydroxyethylamino functionality to make it easier to attach to polymer or dendrimer units. The synthetic routes to the stilbene and diphenylbutadiene series members are illustrated in Figure 1. The synthesis of the N-ethyl-N-hydroxyethyl counterpart is illustrated in Figure 2. Bipolaron formation in these materials is illustrated in Figure 3 by the oxidation of the bis-(diphenylamino)stilbene, which also illustrates the magnitude of the shift in oscillator strength in going from the neutral to the bipolaron form.

SYNTHESIS OF PENDANT POLYMERS, COPOLYMERS AND FUNCTIONALIZED DENDRIMERS

We have previously discussed how bis-(diphenylamino)diphenylpolyene units can be attached to a PMMA backbone[5]. We have utilized this approach as a means of comparing very different chromophores to one another, as the post-polymerization attachment dictates that the polymer MW and polydispersity are the same for each and every case. We have standardized on a 10% attachment for OPL studies in order to insure a high degree of linear transmission through solutions or films (>60%). The pendant polymers are prepared from commercially available poly[methacryloyl chloride] (Monomer and Polymer & Dajac Labs) by adding 0.1 equivalent of the chromophore-OH to the polymer in dioxane or THF. Pyridine (0.1 equiv.) is then added and the polymer solution is heated overnight. A large excess of methanol is then added to convert excess COCl groups into COOMe groups and the resulting solution heated for several hours. The solution is then poured into cold MeOH or water, and the polymer obtained through filtration and drying. Copolymers are obtained by reacting a bis-acyl dichloride with a chromophore-$(OH)_2$ in THF with at least 2-3 equivalents of pyridine. After heating overnight, the copolymer is obtained by pouring into cold methanol, and can be purified further by dissolving in a suitable solvent and reprecipitating in MeOH. Chromophore-functionalized dendrimers are obtained from amino-terminated PAMAM dendrimers (Aldrich Chemical Co.) by an interfacial technique illustrated in Figure 4. In all cases, the polymer UV-Vis spectra is almost identical to the free chromophore (within 10 nm). Oxidation of the polymers in solution with $SbCl_5$ confirmed that the attached or incorporated chromophores form bipolarons to the same extent, and with the same optical properties as the free chromophores.

RESULTS AND DISCUSSION

The use of bis-(diphenylamino)diphenylpolyenes for optical limiting applications has accelerated since the report of Perry, Marder and coworkers[6] that the disubstituted stilbene had an extremely large two-photon cross-section (20 x stilbene itself) for nanosecond pulses. This has been attributed to large transient absorption from the photo-excited states. Sariciftci and Heeger[7] have recently reviewed the use of C_{60} and other electron acceptors to photo-induce charge transfer from conducting polymers (e-donors), forming highly absorbing polaronic species that can be utilized as RSA optical power limiters. Our group has been attempting to utilize similar techniques to photo-induce charge transfer in the bis-(diphenylamino)diphenylpolyenes with 1, 2 and 3 double bonds with a variety of acceptors. The formation of highly absorbing transient species has been observed, but the chemical nature of the species has not yet been identified. We are currently in the process of elucidating the nature of the transient species (polaronic radical cations, bipolaronic dications or neutral triplets), and we hope to be able to report on the detailed mechanism in the near future. What is clear at the present time is that this is an attractive mechanism for optical power limiting in the Visible region.

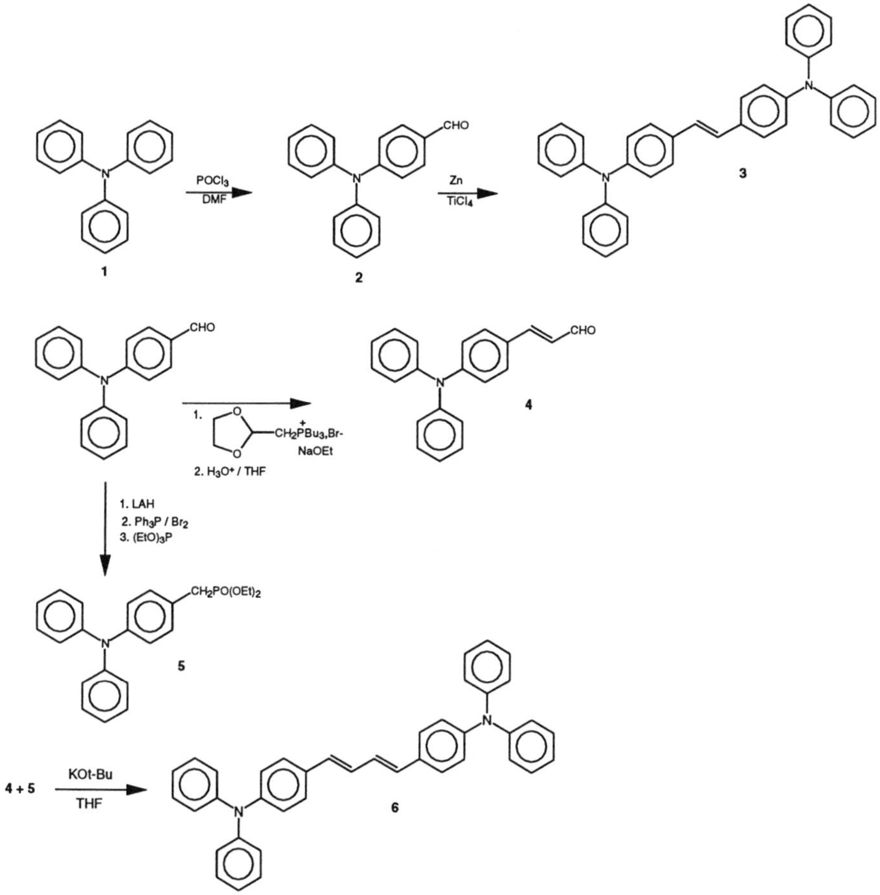

Figure 1. Synthesis of bis-(diphenylamino)diphenylpolyenes (n = 1,2).

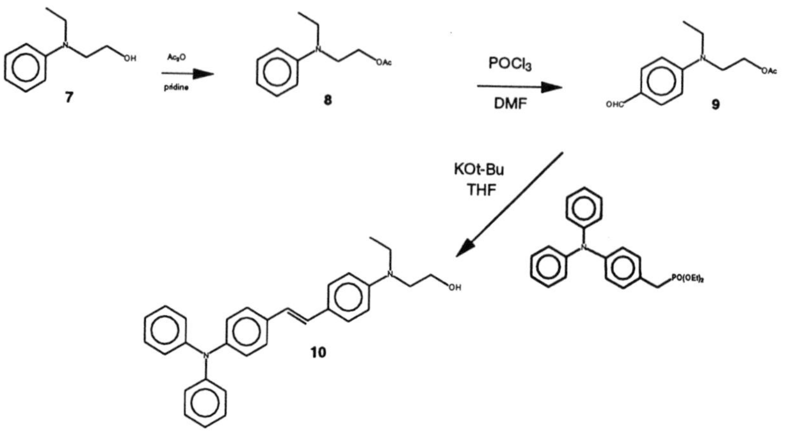

Figure 2. Synthesis of 4-diphenylamino-4'-(N-ethyl, N-2"-hydroxyethyl)stilbene

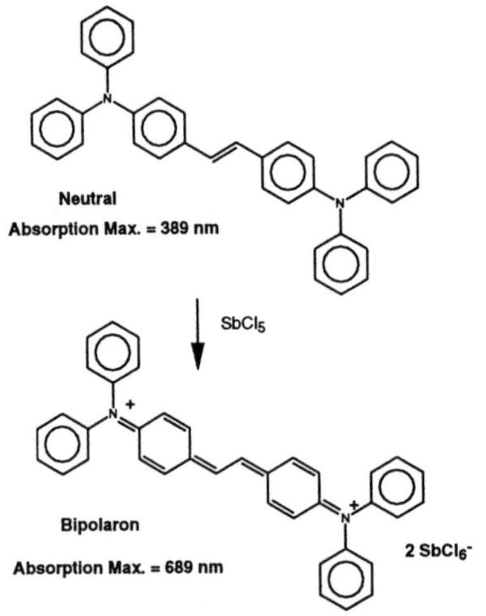

Figure 3. Bipolaron formation in bis-(diphenylamino)-E-stilbene

72

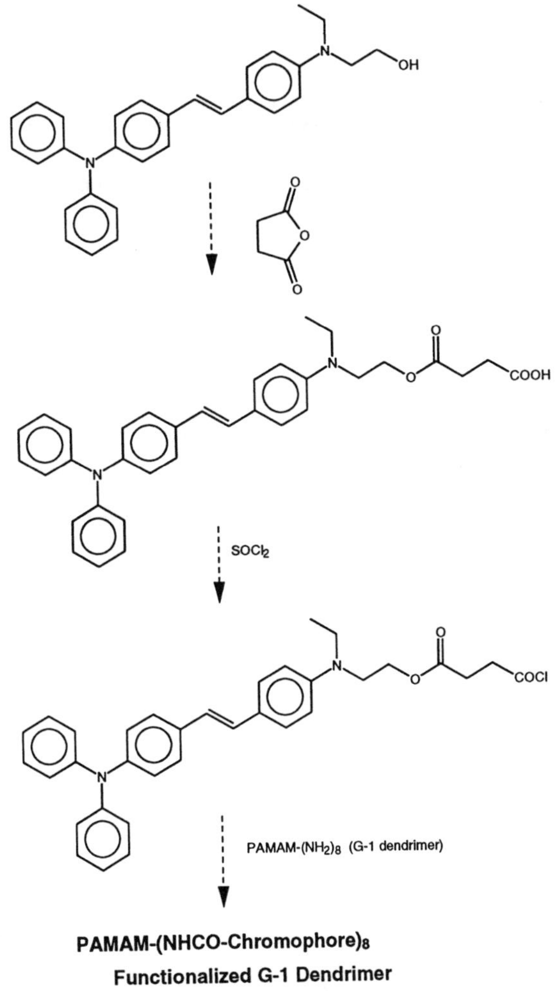

Figure 4. Formation of PAMAM-(NHCO-Chromophore)₈ Functionalized G-1 Dendrimer

As pointed out above, Perry, Marder and coworkers[6] have identified this class of bis-donor polyene as having enhanced two-photon cross sections. Most recently, we learned that preliminary two-photon fluoresence measurement on the bis-(diphenylamino)diphenyltriene and the bis-(diphenylamino)diphenyltetraene showed that these materials also have significant two-photon absorption cross-sections[7]. We are currently pursuing both more detailed two-photon cross-section measurements as well as transient absorption from photo-induced charge-transfer. Since the RSA from photo-induced bipolaron formation will occur at the high energy end of the Visible, and the TPA occurs at the low energy end of the Visible, it is possible that one or more of these bis-(diphenylamino)diphenylpolyenes will convey protection from laser pulses by both mechanisms, thus operating as a bimechanistic optical power limiter.

CONCLUSIONS

We have shown that bis-(diphenylamino)polyenes can be incorporated in a variety of macromolecular structures. Although the pendant polymer approach is the easiest to use when one wishes to compare a large number of different chromophore types, it will probably be difficult to achieve a high percentage loading (> 50%) before the polymer becomes unprocessible. At the current time, we believe that the functionalized dendrimer approach will lead to the greatest chromophore loading. It is also germane to note that dendrimers at the G-3 to G-4 level undergo a conformational reorganization to achieve a globular shape. This will allow the chromophores to be concentrated on the periphery of the structure and thus available for efficient photon interception.

Thus we predict that functionalized dendrimers will be the more efficient two-photon absorber.

ACKNOWLEDGEMENTS

Partial support of this research by the Air Force Office of Scientific Research under grants F49620-96-1-0440 and F49620-97-1-0413 is gratefully acknowledged.

REFERENCES

1. C. W. Spangler, R. K. McCoy, A. A. Dembek, L. Sapochak and B. D. Gates, *J. Chem. Soc. Perkin Trans. 1*, p. 779 (1991).
2. C. W. Spangler, P.-K. Liu, A. A. Dembek and K. O. Havelka, *J. Chem. Soc. Perkin Trans. 2*, p. 1207 (1992).
3. C. W. Spangler and M. Q. He, *J. Chem. Soc. Perkin Trans. 1*, p. 715 (1995).
4. C. Spangler and M. Q. He, in *Handbook of Conductive Molecules and Polymers: Vol. 2. Conductive Polymers: Synthesis and Electrical Properties*, edited by H. S. Nalwa, John Wiley & Sons Ltd., Chichester, 1997, pp. 389-414.
5. C. W. Spangler, T. Faircloth, E. H. Elandaloussi and B. Reeves, *Mat. Res. Soc. Symp. Proc.*, **488**, p. 283 (1998).
6. J. E. Ehrlich, X. L. Wu, I.-Y. S. Lee, A. A. Heikal, Z.-Y. Hu, H. Rockel, S. R. Marder and J. W. Perry, *Mat. Res. Soc. Symp. Proc.* **479**, p. 9 (1997).
7. L. V. Natarajan, private communication.

THE SYNTHESIS OF DIPHENYLPOLENES AND PPV-OLIGOMERS INCORPORATING DIPHENYLPHOSPHINO DONOR GROUPS

L. G. Madrigal*, C.W. Spangler*

*Department of Chemistry and Biochemistry, Montana State University, Bozeman, MT 59717

ABSTRACT

Previous studies of the second and third order nonlinear optical properties of organic molecules have shown enhancement of the nonlinear response when second row elements replace first row elements (e. g. S for O) in electron donating groups. Various substituted amino functionalities have been utilized extensively as strong donors in the design of molecules for both second and third order NLO applications. However, to date, no research groups have reported a systematic study of the effect of replacing N (nitrogen) by P (phosphorus) in various chromophore functionalities. We have now synthesized a series of bis-(diphenylphosphio)diphenylpolyenes with up to five double bonds, a bis-(diphenylphosphino)-PPV dimer, and a diphenylphosphino DANS equivalent. The absorption characteristics compared to the N equivalents, and the implications for photonics applications will be discussed.

INTRODUCTION

Over the past twenty years there have been wide-ranging efforts to increase the nonlinear optical response of organic molecules. Through a large number of detailed structure-property relationships, we now have a firm data base for understanding how changes in the chromophore structure will affect the magnitude of either the second-order hyperpolarizability, β, or the third-order hyperpolarizability, γ. It was recognized fairly early that gross changes such as adding double bonds or increasing the number of aryl rings could cause both β and γ to increase. However, there was not a simple relationship between the formula conjugation length arrived at by counting the number of conjugated double bonds and rings, but rather a more subtle relationship taking into account the degree of twisting out-of-plane as the length increased. A more consistent form of controlling the magnitude of the hyperpolarizability involves the use of substituent effects, wherein groups could add to, or subtract from, the electron density in the main conjugation sequence, and thus altering the HOMO – LUMO gap. This can be illustrated in simple fashion as follows:

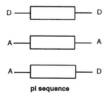

pi sequence

For centrosymmetric molecules, both adding electron density (D,D) or withdrawing e-density (A,A) from the conjugation sequence termini are effective in increasing γ over the unsubstituted system[1]. For non-centrosymmetric chromophores, the magnitude of the second order β is related to the donor and acceptor strength[1] and the degree of bond-length alternation[2] caused by the polarization response. Inherent in all of the above is the recognition that any structural modification of an NLO chromophore which increases either β or γ usually also red-shifts the UV-Vis spectrum, and is usually referred to as the "nonlinearity-transparency" trade-off. In this current study we have undertaken to demonstrate what the effect of replacing a "strong" electron donor, such ad dialkyl or diphenylamino, with the equivalent disubstituted phosphino group. Due to the air sensitivity of dialkylphosphino groups, we have focused on diphenylphosphino comparisons with diphenylamino.

SYNTHESIS OF BIS-(DIPHENYLPHOSPHINO)DIPHENYLPOLYENES

Over the past four years we have extensively studied[3] the synthesis and charge-state (either polaronic radical cations or bipolaronic dications) formation in bis-(diphenylamino)-diphenylpolyenes. This polyene series is interesting from several different perspectives: (1) in general, diphenylpolyenes are very insoluble, particularly the bis-(dimethylamino) series, however, the bis-(diphenylamino) series is very soluble which has allowed us to synthesize a series with up to ten double bonds; (2) upon oxidative doping with $SbCl_5$ in solution, very stable bipolaronic dications are formed, with lifetimes under ambient laboratory conditions of up to several weeks; (3) large red shifts are observed in the spectrum in going from the neutral to bipolaron form, with concomitant large increases in the molar absorptivity. This latter observation has led us to postulate that these compounds might be ideal reverse saturable absorption (RSA) optical power limiters (OPLs) based on photogeneration of these highly absorbing charge states under intense laser irradiation. These materials are currently being evaluated in several different laboratories (both industrial and DoD) for optical power limiting applications. In addition, Perry, Marder and coworker[4] have recently shown that bis-(diphenylamino)stilbene is an excellent two-photon absorbing (TPA) chromophore. We have recently initiated a program of design strategies for new TPA chromophores, and several members of the bis-(diphenylamino)diphenylpolyene series (n = 2,3) have been shown to have large two-photon cross-sections[5].

Synthesis of a bis-(diphenylphosphino)diphenylpolyene series required some synthetic modifications of our usual diphenylpolyene route[6], due to the competition of the highly nucleophilic competition from the P lone e-pair. We can circumvent this competition by complexing the P lone pair with BH_3 as a protecting group. This protecting group can be later removed to yield the free diphenylphosphino groups[7]. The synthetic routes to the n= 1-5 polyene series are illustrated in Figures 1-3.

SYNTHESIS OF BIS-(DIPHENYLPHOSPHINO)-PPV DIMER AND 4-DIPHENYLPHOSPHINO-4'-NITROSTILBENE (DANS EQUIVALENT)

Two other representative compound series of interest in photonics application are oligomers of PPV, and the quintessential second-order NLO chromophore DANS. We have therefore synthesized an α,ω-bis-(diphenylphosphino)-PPV dimer and a diphenylphosphino DANS equivalent. The synthesis of these two interesting materials are shown in Figure 4. Since the bis-

(dimethylamino) counterparts of both of these compounds are highly colored, a comparison of their $\pi - \pi^*$ transition would prove interesting.

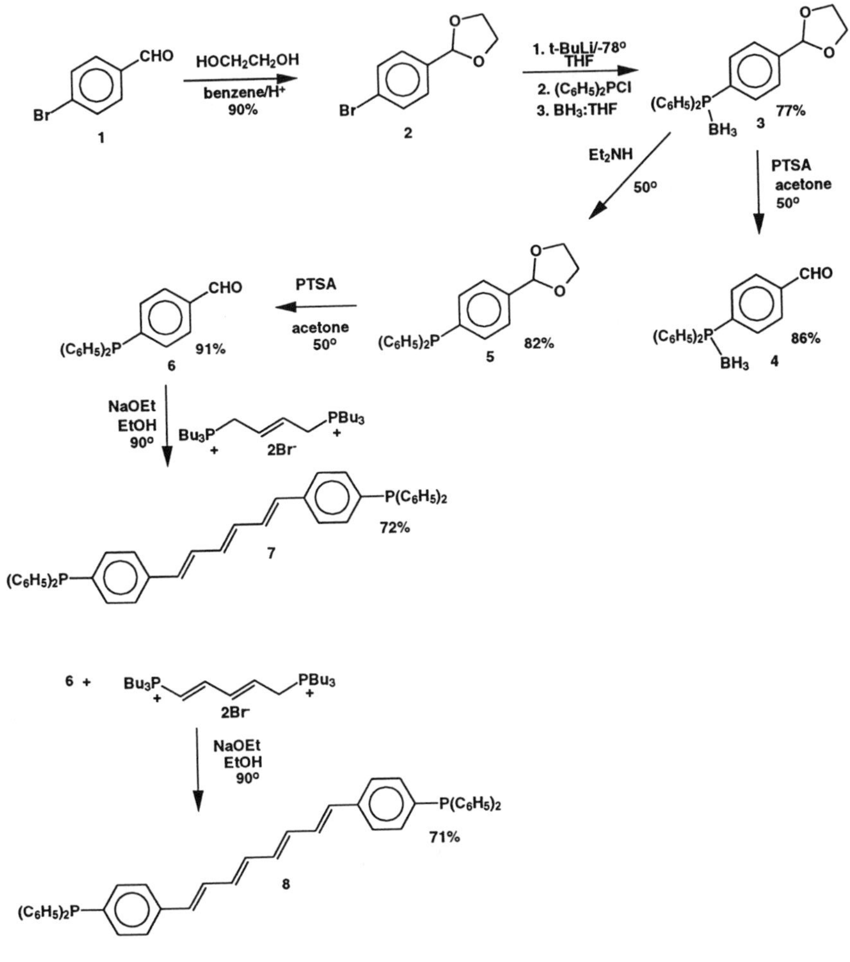

Figure 1. Syntheses of α,ω–bis-(diphenylphosphino)diphenyl-1,3,5-hexatriene abd –1,3,5,7-octatetraene.

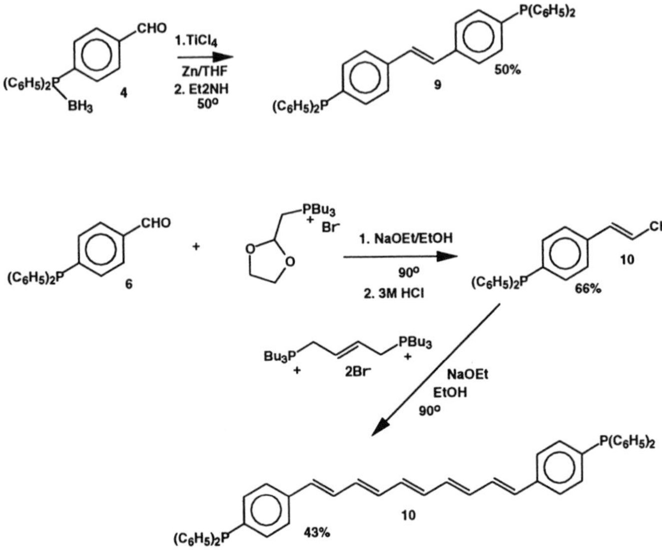

Figure 2. Syntheses of α,ω–bis-(diphenylphosphino)diphenylethene and -diphenyl-1,3,5-decapentaene.

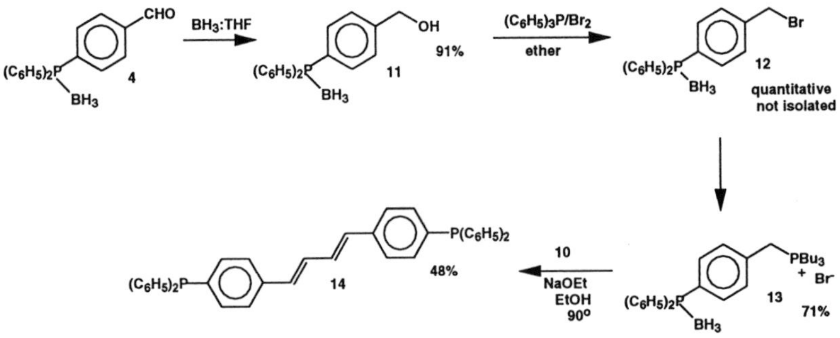

Figure 3. Syntheses of α,ω-bis-(diphenylphosphino)diphenyl-1,3-butadiene.

ABSORPTION SPECTRA OF DIPHENYLPHOSPHINOPOLYENES

We measured the absorption spectra of the bis-(diphenylphosphinodiphenylpolyene series and compared them directly to their bis-(diphenylamino)diphenylpolyene counter parts. The results are displayed in Table 1.

Table 1. Absorption Spectra of Bis-Donor-Diphenylpolyenes

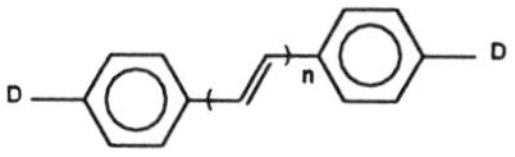

Number of double bonds	Diphenylphosphino λmax	Diphenylamino λmax
1	341 nm	389 nm
2	362	404
3	384	424
4	400	438
5	418	449

As can easily be seen from the data in Table 1, the diphenylphosphino series is blue-shifted, in comparison to the diphenylamino series, by ca. 30 – 40 nm. This remarkable shift may have profound importance in designing photonic materials. As mentionned previously, the nonlinearity-absoptivity trade-off has frustrated many workers in the field, since it did not seem that the two effects could be decoupled. It is important, therefore, that third-order NlO measurements be carried out for the phosphino series, and we hope to be able to report on these measurements in the near future.

Equally intriguing would be the comparison between a D-A polyene compared to a well-known β chromophore. Therefore we synthesized a diphenylphosphino equivalent to DANS. Similarly, with the recent report[4] of large two-photon absorption in bis-(diphenylamino)stilbene and bis-(diphenylamino)-PPV dimers, we also synthesized the bis-(diphenylphosphino)-PPV dimer. The comparison of absorption spectra are shown below:

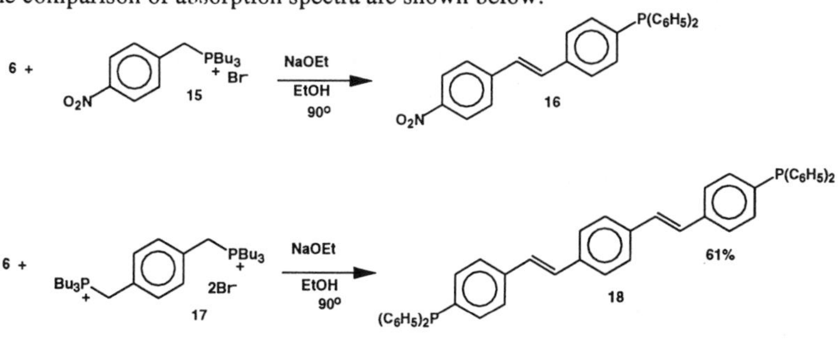

Figure 4. Syntheseof p,p'-bis-(diphenylphosphino)-PPV dimer and 4-diphenylphosphino-4'-nitrostilbene

(C$_6$H$_5$)$_2$P-PPV-dimer λ max = 313 nm

(C$_6$H$_5$)$_2$N-PPV-dimer λ max = 403 nm

λ max = 436 nm   λ max = 364 nm

As can be easily seen again there are dramatic blue shifts in the absorption spectra of the diphenylphosphino compounds. Since blue-shifted chromophores are highly desireable for both second-order NLO materials, and for TPA chromophores for optical limiting in the Visible, both of these chromophores will be tested in the near future for these applications.

CONCLUSIONS

We have shown that compound series wherein P replaces N in donor substituents of interest in various photonics applications can be synthesized. We can anticipate that these materials will be tested for all applications in which their diphenylamino counterparts have excellent properties except for their red-shifted spectra.

If their second- and third-order hyperpolarizabilities prove to be of the same order of magnitude as the diphenylamino series, then these materials should be of great utility for NLO device design. Similarly, if the PPV dimer series proves to have equally large TPA cross-sections compared to the diphenylamino-PPV dimers, then they should have excellent optical limiting capability in the 500 – 700 nm region of the Visible. We will be able to report on all of these possibilities in the very near future.

ACKNOWLEDGEMENTS

Partial supprt of this research by the Air Force Office of Scientific Research under grants F49620-96-1-0440 and F49620-97-1-0413 is gratefully acknowledged.

REFERENCES

1. L.-T. Cheng, W. Tam, S. R. Marder, A. E. Stiegmann, G. Rikken and C. W. Spangler, *J. Phys. Chem.* **95**, p. 10643 (1991).
2. S. R. Marder, D. N. Beratan and L.-T. Cheng, *Science* **252**, p. 400 (!991)
3. C. W. Spangler, T. Faircloth, E. H. Elandaloussi, and B. Reeves, *Mat. Res. Soc. Symp. Proc.* **488**, p. 283 (1998).
4. J. E. Ehrlich, X. L. Wu, I,-Y. S. Lee, A. A. Heikal, Z.-Y. Hu, H. Rockel, S. R. Marder and J. W. Perry, *Mat. Res. Soc. Symp. Proc.* **479**, p. 9 (1997).
5. L. V. Natarajan, Private communication.
6. C. W. Spangler, R. K. McCoy, A. A. Dembek, L. Sapochak, and B. D. Gates, *J. Chem. Soc. Perkin Trans. 1*, p. 186 (1991)
7. T. Imamoto, T. Oshiki, T. Onozawa, T. Kusumoto and K. Sato, *J. Am. Chem. Soc.* **112**, p. 5244 (1990).

SYNTHESIS AND OPTICAL LIMITING CHARACTERIZATION OF A PORPHYRIN-BUCKMINSTERFULLERENE DYAD

KENNETH J. McEWAN [1], KEITH L. LEWIS [1], HOW-GHEE ANG [2], ZHI-HENG LOH [2], LENG-LENG CHNG [2], YIEW-WANG LEE [2]
[1] Defence Evaluation and Research Agency, Malvern, Worcs, WR14 3PS, United Kingdom
[2] DSO National Laboratories, 20 Science Park Drive, Singapore 118230

ABSTRACT

The optical limiting performances of donor-buckminsterfullerene (C_{60}) dyads have hitherto been unreported. The predicted large excited state absorption cross-section of such dyads in their charge-separated states would enable them to exhibit enhanced optical limiting relative to their individual components. We report herein the synthesis and optical limiting characterization of a novel porphyrin-buckminsterfullerene dyad, in which the porphyrin and C_{60} moieties are held within close proximity of one another by an o-phenylene bridge. The porphyrin-C_{60} dyad was synthesized via 1,3-dipolar azomethine ylide cycloaddition, and was obtained in 37.0% yield. Optical limiting measurements at 532 nm using 3-ns pulses show that the limiting performance of the dyad is poorer compared to its model compounds. Possible reasons for the worsened performance of the dyad will be discussed.

INTRODUCTION

Recent coherent anti-Stokes Raman spectroscopy investigations by Lascola and Wright [1] showed that the third-order optical nonlinearities of C_{60} anions are enhanced relative to pristine C_{60}. The enhancement was attributed to a reduction in both the molecular and electronic symmetries of C_{60} upon addition of electrons into its T_{1u} orbital, thereby increasing the number of electronic transition pathways in the excited states of the resultant anion. As both porphyrins [2-5] and C_{60} [6-9] are known to exhibit optical limiting by excited state absorption, the results of Lascola and Wright suggest that the limiting effects of their charged species could be stronger compared to the neutral analogues.

The rich synthetic chemistry of C_{60} [10,11], along with its electron-deficient nature [12], has enabled the molecule to be functionalized with an array of electron donors [13,14]. In this study, we synthesized a porphyrin-C_{60} dyad (see figure 1) in which the porphyrin and C_{60} moieties are held within close proximity of one another by an o-phenylene bridge. The small through-space separation between the porphyrin and C_{60} moieties in the dyad would enable us to utilize photoinduced electron transfer as a means of generating the charge-separated state. The optical limiting responses of the resultant porphyrin monocation and C_{60} monoanion could then be observed. The results from the optical limiting measurements performed on the dyad and its model compounds will be presented.

EXPERIMENT

The porphyrin-C_{60} dyad was synthesized via 1,3-dipolar azomethine ylide cycloaddition [15] (see figure 2). A yield of 37.0% was obtained, and the product was characterized by UV-vis, steady-state fluorescence and one- and two-dimensional nuclear magnetic resonance (NMR) spectroscopy methods.

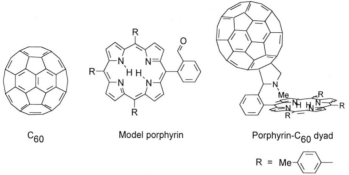

C_{60} Model porphyrin   Porphyrin-C_{60} dyad

Figure 1. Molecular structures of the porphyrin-C_{60} dyad and its model compounds.

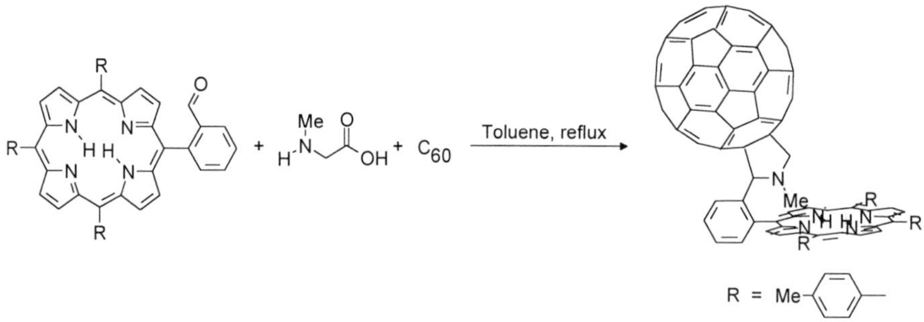

Figure 2. Synthetic route for the preparation of the porphyrin-C_{60} dyad.

For optical limiting measurements, the output from a Q-switched Continuum Minilite Nd:YAG laser was frequency-doubled to generate 532 nm pulses of 3-ns pulse widths. Measurements were performed in the single shot mode on solutions with linear transmittances of ~40%. The input energy ranged from 0.010 – 500 µJ. No signs of damage were observed in the samples even after irradiation at the maximum beam energy.

Geometry optimization of the dyad was carried out in Spartan 5.1 [16]. The structure was pre-optimized using the MMFF94 molecular mechanics force field [17], followed by refinement using the PM3 Hamiltonian [18] at the restricted Hartree-Fock level. ^{1}H-^{1}H nuclear Overhauser effect spectroscopy was used to verify the chemical accuracy of the optimized structure.

RESULTS AND DISCUSSION

Inter-Chromophore Interaction

The UV-vis absorption maxima of the dyad show negligible shifts relative to its model compounds [19,20], thus suggesting weak through-space interaction between the porphyrin and C_{60} moieties in the ground state. The insignificant spectral perturbation is in contrast to the large bathochromic shifts in the Soret and Q-bands reported previously for several porphyrin-C_{60} dyads [21-24].

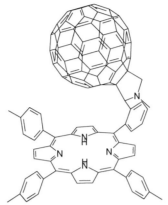

Figure 3. Optimized structure of the porphyrin-C_{60} dyad obtained from molecular modeling.

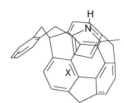

Figure 4. Z-clipped structure of the dyad showing the geometry of inter-chromophore overlap. The fullerene C-C double bond that has the shortest distance of approach to the pyrrole plane is marked with an 'X'.

Molecular modeling of the dyad yielded a structure (see figure 3) with an inter-chromophore π-π separation of 3.28 Å. It is noteworthy that this distance is within the limit for van der Waals overlap between the two chromophores. Moreover, the calculated center-to-center distance of 6.13 Å is shorter than those reported for similar dyads which exhibit strong inter-chromophore interaction [21,22]. Hence, the absence of any through-space interaction in the dyad is clearly unexpected.

Closer examination of the structure (see figure 4), however, shows that the inter-chromophore approach is between a pyrrole sub-unit residing on the porphyrin plane and a fullerene C-C double bond (marked with an 'X' in figure 4). The absence of any perturbation in the UV-vis spectrum of the dyad could therefore be explained by an observation made by Hunter and Sanders [25]. They found that it is the electron-richness or electron-deficiency of the molecular fragments at the region of inter-chromophore contact that control the strength of interaction rather than the redox potentials of the chromophores. Hence, the disposition of the electron-rich pyrrole sub-unit and fullerene C-C double bond towards each other, as dictated by the structural rigidity of the *o*-phenylene bridge, results in π-π electron repulsion instead of van der Waals through-space interaction. This explains for the weak through-space inter-chromophore interaction observed in the dyad.

Optical Limiting Performance

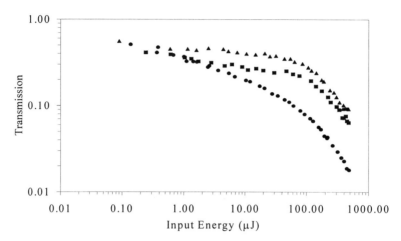

Figure 5. Logarithmic plots of transmission against input energy for C_{60} (circles), model porphyrin (squares) and porphyrin-C_{60} dyad (triangles)

Results obtained from the optical limiting measurements on the dyad, model porphyrin, and C_{60} are shown in figure 5. The saturation threshold of the dyad is approximately 4 times higher than that of C_{60} and 1.3 times higher compared to the model porphyrin. Several reasons have been proposed to account for the poorer limiting performance of the dyad.

First, a charge-separated lifetime in the regime of tens of picoseconds has been reported in the literature for an analogous porphyrin-C_{60} dyad [22]. Hence, even if the dyad in the charge-separated state exhibits enhanced limiting relative to its model compounds, the 3-ns pulse width used in our optical limiting measurements is clearly too wide to allow the observation of any enhancement effects. The presence of any enhancement, if any, can only be observed under picosecond irradiation. Characterization of the optical limiting performance of the dyad under picosecond irradiation will be carried out in the near future.

Second, quenching of the singlet states of both porphyrin and C_{60} moieties of the dyad, which is due to photoinduced electron transfer, can been observed from the steady-state fluorescence spectrum of the dyad. Furthermore, optical limiting by porphyrins and C_{60} under nanosecond irradiation is due to excited state absorption originating from the triplet state. Since photoinduced electron transfer competes with intersystem crossing as an alternative decay pathway for the singlet state, the reduced quantum yield of the triplet state therefore results in poorer optical limiting performance of the dyad. A similar explanation has been proposed recently by Sun *et al.* to account for the weak optical limiting effect of several aminofullerenes derivatives [26].

Third, the excited state absorption spectra of the dyad might be shifted relative to its model compounds. Such a shift could be accompanied by a decrease in the excited state absorption cross-section of the dyad at the probe wavelength. The resultant decrease in the figure-of-merit for optical limiting would explain for the poorer limiting performance of the dyad.

Finally, the chemical functionalization of C_{60} and porphyrin reduces the molecular symmetry of the dyad with respect to its model compounds. The detrimental effect of reduction of molecular symmetry on the optical limiting performance of the resultant C_{60} derivative, as observed in previous studies [27-30], could therefore explain for the poorer limiting performance of the dyad compared to its model compounds.

CONCLUSIONS

A porphyrin-C_{60} dyad, in which the close proximity of the porphyrin and fullerene moieties was achieved *via* the use of an *o*-phenylene bridge, has been synthesized. Electronic spectroscopy shows the absence of any significant through-space inter-chromophore interaction in the dyad. This observation could be rationalized by molecular modeling, which shows the unfavorable disposition of the electron-rich bonds on each chromophore towards one another.

Optical limiting measurements at 532 nm using 3-ns pulses showed that the saturation threshold of the dyad is approximately 4 times higher compared to C_{60} and 1.3 times higher compared to the model porphyrin. This could be due to an ultra-short lifetime for the charge-separated species, depopulation of the triplet state by electron transfer, a shift in the excited state absorption spectra of the dyad relative to its model compounds, or a decrease in the figure-of-merit of the dyad upon chemical functionalization. These postulates await confirmation from time-resolved laser spectroscopy experiments, which are currently in progress.

The rich synthetic chemistry has enabled C_{60} to be functionalized with an array of electron donors. Further work should be directed towards the synthesis of donor-C_{60} dyads with long charge-separation lifetimes to enable observation of their optical limiting performance under nanosecond irradiation.

ACKNOWLEDGMENTS

We thank the Department of Chemistry, National University of Singapore, for technical support on NMR characterization. Joint funding from DSO National Laboratories and Directorate of Research and Development is acknowledged.

REFERENCES

1. R. Lascola and J.C. Wright, Chem. Phys. Lett. **269**, 79 (1997); **290**, 117 (1998).

2. W. Blau, H. Bryne, W.M. Dennis, J.M. Kelly, Opt. Commun. **56**, 25 (1985).

3. S. Guha, W.T. Roberts, B.H. Ahn, Appl. Phys. Lett. **68**, 3686 (1996).

4. P. Chen, I.V. Tomov, A.S. Dvorniknov, M. Nakashima, J.F. Roach, D.M. Alabran, P.M. Rentzepis, J. Phys. Chem. **100**, 17507 (1996).

5. W. Su, T.M. Cooper, M.C. Brant, Chem. Mater. **10**, 1212 (1998).

6. L.W. Tutt and A. Kost, Nature **356**, 225 (1992).

7. D.G. McLean, R.L. Sutherland, M.C. Brant, D.M. Brandelik, P.A. Fleitz, T. Pottenger, Opt. Lett. **18**, 858 (1993).

8. B.L. Justus, Z.H. Kafafi, A.L. Huston, Opt. Lett. **18**, 1603 (1993).

9. S.R. Mishra, H.S. Rawat, S.C. Mehendale, Appl. Phys. Lett. **71**, 46 (1997).

10. A. Hirsh, *The Chemistry of the Fullerenes*, (Georg-Thieme, Stuttgart, 1994).

11. *The Chemistry of Fullerenes*, edited by R. Taylor (World Scientific, Singapore, 1995).

12. R.C. Haddon, Science **261**, 1545 (1993).

13. H. Imahori and Y. Sakata, Adv. Mater. **9**, 537 (1997).

14. N. Martín, L. Sánchez, B. Illescas, I. Pérez, Chem. Rev. **98**, 2527 (1998).

15. M. Prato and M. Maggini, Acc. Chem. Res. **31**, 519 (1998).

16. Spartan 5.1, Wavefunction, Inc., 18401 Von Karman Ave., Suite 370, Irvine, CA 92612, U.S.A.

17. M. Clark, R.D. Cramer III, N. van Opdensch, J. Comput. Chem. **10**, 982 (1989).

18. J.J.P. Stewart, J. Comput. Chem. **10**, 209, 221 (1989).

19. UV-vis λ_{max}/nm (model porphyrin/CH_2Cl_2) = 423, 518, 553, 593, 650; λ_{max}/nm (dyad/CH_2Cl_2) = 228, 256, 423, 518, 553, 593, 654. The high energy excitation peaks at 228 and 256 nm in the UV-vis spectrum of the dyad originate from the C_{60} moiety, and are unperturbed relative to the pristine fullerene [20].

20. S. Leach, M. Vervolet, A. Desprès, E. Bréheret, J.P. Hare, T.J. Dennis, H.W. Kroto, R. Taylor, D.R.M. Walton, Chem. Phys. **160**, 451 (1992).

21. P.A. Liddell, J.P. Sumida, A.N. MacPherson, L. Noss, G.R. Seely, K.N. Clark, A.L. Moore, T.A. Moore, D. Gust, Photochem. Photobiol. **60**, 537 (1994).

22. D. Kuciaskas, S. Lin, G.R. Seely, A.L. Moore, T.A. Moore, D. Gust, T. Drovetskaya, C.A. Reed, P.D.W. Boyd, J. Phys. Chem. **100**, 15926 (1996).

23. H. Imahori, K. Hagiwara, M. Aoki, T. Akiyama, S. Tamiguchi, T. Okada, M. Shirakawa, Y. Sakata, J. Am. Chem. Soc. **118**, 11771 (1996).

24. P.S. Baran, R.R. Monaco, A.U. Khan, D.I. Schuster, S.R. Wilson, J. Am. Chem. Soc. **119**, 8363 (1997).

25. C.A. Hunter and J.K.M. Sanders, J. Am. Chem. Soc. **112**, 5525 (1990).

26. Y.-P. Sun, B. Ma, C.E. Bunker, J. Phys. Chem. A **102**, 7580 (1998).

27. Y.-P. Sun, J.E. Riggs, B. Liu, Chem. Mater. **9**, 1268 (1997).

28. Y.-P. Sun, G.E. Lawson, J.E. Riggs, B. Ma, N. Wang, D.K. Moton, J. Phys. Chem. A **102**, 5520 (1998).

29. R. Signorini, M. Zerbetto, M. Meneghetti, R. Bozio, M. Maggini, G. Scorrano, M. Prato, G. Brusatin, E. Menegazzo, M. Guglielmi, SPIE Fullerenes and Photonics III **2854**, 130 (1996).

30. M. Meneghetti, R. Signorini, M. Zerbetto, R. Bozio, M. Maggini, G. Scorrano, M. Prato, G. Brusatin, E. Menegazzo, M. Guglielmi, Synth. Met. **86**, 2353 (1997).

ELECTRICAL FIELD EFFECTS ON CRYSTALLINE PERFECTION OF MBA-NP CRYSTALS BY MAPPING OF BRAGG-SURFACE DIFFRACTION

S.L. MORELHÃO*, L.H. AVANCI**, M.A. HAYASHI**, L.P. CARDOSO**
*Depto of Applied Physics, University of São Paulo, São Paulo, Brazil, morelhao@if.usp.br
**Institute of Physics, Univesity of Campinas, Campinas, Brazil

ABSTRACT

The interaction of weak electrical field with the molecular dipole of the MBA-NP crystals is investigated. Such interaction affects the misorientation and size of the perfect diffracting regions of the crystal, which is monitored by mapping the Bragg-surface diffraction condition. This diffraction technique has also confirmed that the diffracting regions are large enough to allow primary extinction and the non-uniform mosaicity along the crystal surface.

INTRODUCTION

High quality non-centrosymmetric organic single crystals, such as 2-(α-methylbenzylamino)-5-nitropyridine (MBA-NP), have exceptionally large second order polarisabilities. The possibilities of exploiting this fundamental property are perceived to be extensive because of the almost endless variations in chemical structure that can be produced through modern organic synthesis. In addition to the exceptional nonlinearities, all space groups are of the type that also produces piezoelectricity.

In terms of controlling material nonlinear properties, characterization of the piezoelectric tensor has been important [1]. X-ray diffraction techniques are often used to determine small changes in the crystal lattice under an applied electric field. Recently, a method based on the X-ray multiple beam diffraction phenomenon was developed for determining more than one piezoelectric coefficients from a single ϕ scan (azimuthal scan) [2]. A strong limitation for applying the method in MBA-NP crystals is the unexpected features often observed in the scans, such as peak broadening, splitting and shifting in magnitude with can not be correlated to any piezoelectric property of the material. In order to properly address this issue, the source of these features, we have used an X-ray diffraction technique that consist in mapping the solid angle around a three-beam diffraction condition [3].

THEORY

The three beam diffraction, as well as any multiple beam diffraction, is different from conventional two beam diffraction because it defines a specific direction, not a cone of direction, to be fulfilled by the wavevector of the incident beam. A special three-beam diffraction, called Bragg-surface diffraction (BSD) [4], was chosen to be investigated in this work. It presents a strong signal and a short extinction distance for the secondary surface reflection [5]. The incident wavevector direction for the BSD is illustrated in Fig. 1 by the vector TO, from the T-point to the origin O of the reciprocal space of a perfect diffracting region. And, the position of the T-point is given by the intersection of three spheres (dispersion surfaces) centered at the origin O and, at the extreme of the $\mathbf{H}_{01}$ and $\mathbf{H}_{02}$ diffraction vectors. The 01 and 02 reflections are the symmetric Bragg reflection and the surface reflection, respectively. The $\mathbf{H}_{21}$ diffraction vector stands for the coupling reflection, which is responsible for the energy transfer between the diffracted beams. This energy transfer process changes the intensities, and by monitoring the Bragg diffracted beam, the beam from the 01 reflection, the occurrence of BSD can be measured.

The perfect regions must be large enough, of the order of the extinction distance for the surface reflection, to allow primary extinction regime of energy transfer. Only under this regime, the T-point exists for each perfect region [2]. Otherwise, the whole crystal defines the T-point as the convolution between the mosaic spreads of the 02 (surface) and 21 (coupling) reflections. That is the secondary extinction regime, the dominant regime for ideally imperfect crystals.

In a crystal made up of large perfect regions, the BSD condition is given by an aggregate of T-points, Fig. 2. When the solid angle around this aggregate is mapped by combining ω (incidence angle) and ϕ (azimuthal rotation) scans, the spatial distribution of

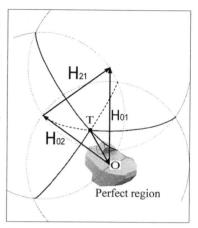

Fig. 1: Bragg-surface diffraction condition for a perfect diffraction region. The incedent wavevector direction is given by the TO vector.

perfect region misorientation, i.e. the mosaic distribution, is visualized. For mosaic crystals where the secondary extinction dominates, the BSD profiles are gaussians in the ω scan, as well as in the ϕ scan [3]. Moreover, it is important to emphasize that the size of the perfect region, at the surface diffracted beam direction, is the region dimension defining the regime in which the BSD does occur.

EXPERIMENT

The molecular dipole of the MBA-NP molecules is formed between the nitrogens in the amino and nitro groups, and the resultant moment of dipole p is along the b lattice vector of the unit cell formed by two molecules [6]. Lattice parameters of the unit cell are $a=5.408$Å, $b=6.371$Å, $c=17.968$Å, $\alpha=\gamma=90°$, and $\beta=94.60°$. The (001) plane is a natural cleavage plane of crystal. The analyzed sample was cut in dimensions of about $10\times10\times3$mm^3, with the largest face as the (001) plane, and the lateral faces as the (100) and (010) planes. The electrical field was applied in the crystal through conductive sponges, placed between metal plates and the sample lateral faces.

The measurements were carried out at the synchrotron radiation source (SRS), high-resolution diffraction station 16.3 [7], Daresbury Laboratory, Warrington, U.K.. The wiggler beam was monochromatized to 8.334 keV by two Si (111) channel-cut crystals, and limited by 0.1×0.1 mm^2 cross slit screen. It provides an incident beam with horizontal and vertical divergences smaller than 5 arcsec. The linear

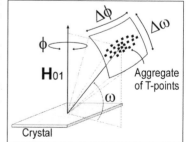

Fig. 2: An aggregate of T-points defines the BSD condition for a mosaic crystal with large diffracting regions.

polarization of the synchrotron beam was perpendicular to the incidence plane of the 01 reflection. BSD mapping were performed with step size of 9 arcsec in both ω and ϕ axes, and counting time of 0.5 seconds.

A Si (111) channel-cut crystal was used as analyzer for reciprocal space mapping (RSM) measurements. The high resolution of the diffractometer at station 16.3 allows the coupled scan of the ω and 2θ (detector arm) axes. These measurements give the intensity distribution as function of the angles 2θ and ξ, where $\xi=\omega-2\theta/2$. Step size of 18 arcsec was used in the ω axis, and the detector axis coupled, in order to provide the same step size in ξ.

RESULTS

The BSD chosen for our measurements has the $00\bar{6}$ (Bragg) and $\bar{1}0\bar{3}$ (surface) reflections.

Fig. 3: BSD mapping performed at two adjacent areas near the center of the sample surface. The angular axes are in arcsec.

Their structure factors are 9% and 31% of the one for the $1\bar{1}\bar{3}$ reflection, respectively. This BSD is observed at around $\omega_0=14.431°$ and $\phi_0=1.513°$, if the $[0\bar{1}0]$ direction is taken as the reference direction ($\phi=0$). The first interesting result is the changes in the BSD mapping as a function of the illuminated area in the sample. In Fig. 3, two maps performed with the incident beam hitting different areas near the center of the (001) face are presented. The map in Fig. 5a is also from a different area. The background intensity due to the Bragg diffraction was removed, by first measuring its profile in a ω scan away from the BSD position, and then subtracting it from the maps.

In order to analyze the mosaicity of the crystal with a standard X-ray diffraction technique, RSM of the Bragg reflection was also carried out at the station 16.3. The results from RSM performed at two ϕ position are shown in Fig. 4, one with the *a* axis in the incidence plane ($\phi=90°$) and the other with the *b* axis in this plane ($\phi=0$). Due to the incidence geometry and the relatively deep penetration of the X-rays in the sample, the diffracting volumes at each of these measurements are not the same, even if the illuminated area was maintained during ϕ rotation. Although, the two maps are different, in both the mapped reflection is much wider along the ξ

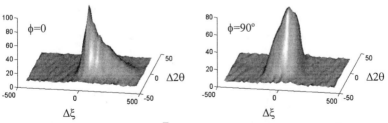

Fig. 4: Reciprocal space mapping of the $00\bar{6}$ Bragg reflection performed at two different ϕ positions. $\Delta\xi$ and $\Delta2\theta$ are in arcsec.

axis, and the peak broadening along the 2θ axis is the same, of the order of 50 arcsec. The effects of the electric field onto the BSD mapping were verified by the following procedure. After fixing the electric contacts and aligning the sample (H_{01} parallel to the φ axis), one mapping was carried out. Then, the voltage supplier was turned on at 100V, and another mapping was performed exactly at the same area. With the electrical field parallel to the polar axis of the crystal, axis b, not even a small change has been observed. On the other hand, when the field was perpendicular to the polar axis, along the [100] direction (a axis), drastic changes can be seen. In Fig. 5, it is shown the BSD mapping before and after the electric field be applied. The mosaic distribution before (Fig. 5a) presents a major peak spread over 100 arcsec, with a minor preferential alignment, for Δω>0, along the diagonal of the mapped angular range. Under the field (Fig. 5b), the peak seems to split up into several peaks, distributed over more than 500 arcsec, and the preferential alignment at ~45° with the φ axis is much clear. The roughness in this map results from the background removing process above mentioned. The ω scans of the background intensities are shown in Fig. 6. The first distribution (in Fig. 5a) is not recovered by turning off the voltage or even changing its signal. Due to our simple set up for applying the field, it could not be re-applied in the [010] direction without compromise the position where the beam hits the sample.

DISCUSSION

The aspect of the BSD mapping in Fig. 3 demonstrates that the BSD condition for the MBA-NP crystal is defined by an aggregate of T-points. Therefore, it falls in a crystal model made up of perfect regions large enough (in the surface parallel direction) to undergo primary extinction. The absence of a low intensity gaussian profile around the BSD condition does eliminate the occurrence of secondary extinction (i.e. rescattering among the regions), and also the existence of some significant amount of small regions. Theoretically, the primary extinction distance for the surface reflection is null, because it is an extremely asymmetric reflection with direction cosine zero. In practice, even diffracting regions much smaller than 1 μm should define a T-point. But in low absorption crystals (like organic crystals), such small regions would give rise to secondary extinction. Laminar diffracting regions, thin in thickness and large in the in-plane direction, provide an appropriated crystal model for the MBA-NP. In this case, the BSD mapping gives the mosaic distribution mainly from stacked regions than from adjacent ones. The observed changes in the maps due to crystal translation, suggest that the in-plane size of the regions should be smaller than the incident beam size at the sample surface (~0.15×0.60mm²), or smaller than 100μm.

The RSM measurements allow just to visualize the projection, into the incidence plane, of the misorientation of the H_{01} diffraction vectors for the diffracting regions, regardless their in-plane size. The reuslts in Fig. 4 are very characteristic of crystals with mosaic distribution, in a magnitude of the distribution observed in the BSD mapping, 100 arcsec. The spatial non-symmetric mosaic distribution are responsible for the changes seen in the RSM performed at φ=0 and φ=90°. The thickness of the diffracting regions is perhaps the most important information that we withdraw from RSM. For instance, the peak width in the 2θ direction of 50 arcsec does imply in a thickness of about 0.6 μm.

Finally, let us discuss the electric field effects into the mosaic distribution. Each perfect region has a total dipole moment, which is the number N of unit cells in the region times the resultant dipole **p** of the unit cell. In other words, each region is a domain with total dipole moment **Np**. Under an electric field **E**, the potential energy of each domain is -N**E·p**, which is proportional to their volume. It is minimum when the field is parallel to the polar axis of the crystal, and maximum when they are anti-parallel in an unstable equilibrium position. In both configurations the torque over the region is zero. Therefore, the mosaic distribution should not be affected by the field along the polar axis, as observed. On the other hand, when **E** and **p** are perpendicular, the torque is maximum over the regions, and they can still realize some work but not greater than NEp. This potential energy can rotate as well as break a large region in smaller ones, all depends on the regions boundary constrains and on their internal bonding energies.

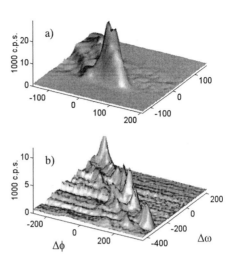

Fig. 5: BSD mapping performed at the same sample position. a) Before and b) after the electric field be stablished along the [100] direction. $\Delta\omega$ and $\Delta\phi$ are in arcsec.

In Fig. 5a, the minor preferential alignment falls over the ω and ϕ positions in which the H_{02} diffraction vectors are touching the Ewald sphere. Then, it is the track of the secondary reflection. For this BSD mapping, the diffracting volume presents some amount of small regions, as for exemple, scratch in the surface or a deeper imperfection. They do not define a BSD peak, but they are able to rescatter the secondary beam when its direction is below the surface, i.e. for $\Delta\omega>0$.

Under the $E \perp p$ configuration, an in-plane, or azimuthal, rotation of the regions would split up the distribution seen in Fig. 5a into another distribution where the T-points were aligned along the ϕ axis and, depending on the crystalline perfection, their respective secondary reflection track would be visible. In Fig. 5b two tracks are clearly seen. Note that they are not continuous, and they are determined by well-defined low-intensity peaks. It shows that the mosaic distribution, or aggregate of T-points, in Fig. 5a split up in only two major regions (or T-points). The other regions in the distribution are not large enough to generate T-points, but they are rescattering the secondary beams of these two major regions, and so the tracks are seen. A few of these regions exist since the tracks are not continuous and the secondary extinction regime of energy transfer among them did not define a gaussian T-point.

Due to the length of the tracks we can not determinie the peaks that are the T-points, or even if they are in the mapped range. However, by analyzing the ω-scan of the background intensity in Fig. 6b, we can figure out the spatial misorientation of these T-points. The scan presents four peaks, (1), (2), (3), and (4), and by just analyzing this scan, we would say that there are four diffracting regions. The BSD mapping shows that the peaks (3) and (4), separated by $\Delta\omega_{track} = 70$ arcsec, are just the tracks from the two major regions, responsible for the peaks (1) and (2). The out-of-plane and in-plane misalignments of these regions are $\Delta\omega_T = 150$ arcsec and $\Delta\phi_T = (\Delta\omega_T - \Delta\omega_{track})/\tan(\sim45°) = 80$ arcsec, respectively.

The boundary constrains avoid a pure in-plane rotation, but the weak bounding between the (001) planes allows the observed out-of-plane rotation of the regions. In the out-of-plane rotation, the potential energy is not reduced. Then, it should not occur, unless to overcome some constrains and allows the in-plane rotation.

CONCLUSIONS

In this work, we have demonstrated that for certain directions, even a weak and static electric field can increase the mosaic distribution of MBA-NP crystals. It does generate features that are able to jeopardize the data analysis from conventional X-ray diffraction techniques such as ω scans, ω/2θ scans, or even φ scans. The BSD mapping was successfully applied for crystalline perfection characterization in this material. It has confirmed that the crystal presents several large regions, and also that BSD mapping allows the spatial visualization of the mosaic distribution.

ACKNOWLEDGMENT

The authors acknowledge Prof. John N. Sherwood for the MBA-NP crystal and Dr. Steve Collins for valuable help at SRS, station 16.3. Financial support from Brazilian agencies CNPq and FAPESP are also acknowledged.

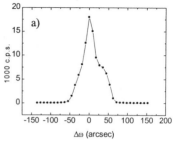

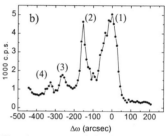

Fig. 6: ω-scans of the background intensities removed from the BSD mapping in Fig. 5. a) from Fig. 5a and b) from Fig. 5b.

REFERENCES

1. P.J. Halfpenny, J.N. Sherwood, and G.S. Simpson in *Orgnic Non-Linear Optical Materials: Crystal Growth and Structural Characterization*, Kikan Kagaku (Quartely Review of Chemistry – Chemical Society of Japan) **15**, 95-121 (1992).
2. L.H. Avanci, L.P. Cardoso, S.E. Girdwood, D. Pugh, J.N. Sherwood, K.J. Roberts, *Physical Review Letters* **81** (24), 5426 (1998).
3. S.L. Morelhão, L.P. Cardoso, *Journal of Applied Crystallography* **29**, 446 (1996).
4. M.A. Hayashi, S.L. Morelhão, L.H. Avanci, L.P. Cardoso, J.M. Sasaki, L.C. Kretly, S.L. Chang, *Applied Physics Letters* **71**(18), 2614 (1997).
5. Surface reflections are those with diffracted beam in the surface parallel direction.
6. A. Carenco, J. Jerphangnon, A. Perignd, *Journal of Chemical Physics* **66**, 3806 (1997).
7. S.P. Collins, R.J. Cernik, C.C. Tang, N.W. Harris, M.C. Miller, G. Oszlanyi, *Journal of Synchrotron Radiation* **5**, 1263 (1998).

SYNTHESIS AND CHARACTERIZATION OF SOME NEW METAL-CONTAINING POLYMERS FOR THIRD ORDER NONLINEAR OPTICAL APPLICATIONS

MAMOUN M. BADER*, PHUONG-TRUC T. PHAM, ANGELA MOLLI, JUSTIN MOSER, KEVIN MEYRES, LESLIE ALBRIGHT
Department of Chemistry, The Pennsylvania State University, Hazleton, PA 18201.

ABSTRACT

We report here our preliminary results on the synthesis and characterization of a new family of metal-containing polymers incorporating 8-hydroxyquinoline. These polymers have the following structural features: a metal center, a chromophore, and a flexible spacer group with variable length. These structural features allow us to have an appreciable degree of control over various physical properties of these materials. We report on three different types of polymers: polyethers, polyesters, and polyamides. The chromophores in these polymers were incorporated in the backbone and in one case as a side chain of polyesters.

INTRODUCTION

Metal-containing or coordination polymers have increasingly been the subject of renewed scientific attention. [1] There have been scattered reports in the literature on coordination polymers since the early 1960's [2]. Earlier reports however were directed at the potential application of bis bidentate ligands capable of formation of such polymers in analytical chemistry [3]. When coordination takes place between a metal ion and an organic ligand, the properties of both are changed in a variety of ways. In most cases the organic ligand is stabilized towards hydrolysis, chemical reagents and toward high temperatures. For instance, it was shown that the hydrolysis of a Schiff base is much more rapid than that of its metal chelates [4]. This type of enhanced stability could find use in the field of liquid crystalline materials containing this functionality. We describe herein our efforts to use such anticipated enhanced stability in the preparation of a new class of metal containing, or coordination polymers. Our ligand of choice was the well-known chelating ligand 8-hydroxyquinoline shown in Figure 1.

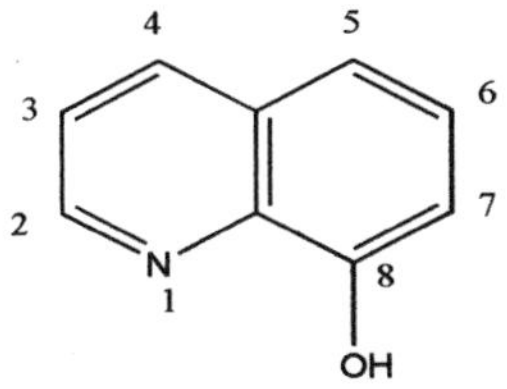

Figure 1: Structure and Numbering System of 8-Hydroxyquinoline

There are several reasons for choosing 8-hydroxyquinoline: one is that we have performed theoretical studies on the nonlinear optical properties of this molecule and several of its derivatives and concluded that, materials containing this structural unit would have interesting second- and third- order nonlinear optical properties [5]. In addition, this ligand is known to be able to form very stable complexes with almost all metals except the alkali metals, and therefore giving us the opportunity to systematically change the metal center and examine its effect on the various physical properties of interest. The fact that the compound aluminum tris(8-hydroxyquinoline), or Alq3, is a prototype electroluminescent material adds to our interest in this molecule [6]. We would like to synthesize a series of these polymers and examine the optical and the electroluminescent properties of these materials. We like to emphasize the large scope of this study and the wide range of possible structures accessible through this methodology.

EXPERIMENTAL

All chemicals were purchased from Aldrich Chemical Company. p-Aminobenzoic acid was crystallized from water, p-nitroaniline was crystallized from ethanol before use. All other chemicals were used as received. Our synthetic strategy involved the following reactions: diazo coupling between p-amino-diazonium chloride and 8-hydroxyquinoline to give compound **1**. The acid chloride, which was obtained by treating compound **1** with thionyl chloride, was then reacted with α,ω-diols and α,ω-diamines to give the diester and the diamide bis bidentate ligands respectively. These ligands were then used to prepare the coordination polymers by reacting them with the divalent metal salt (usually the acetate) in hot ethanol. The diethers were prepared according to literature methods, starting with the reaction of p-nitrophenol with the α,ω-dibroalkanes [7]. The resulting α,ω-p-nitrophenoxy alkanes were then reduced using Sn/HCl to yield the diamines. The diamines were used to prepare the corresponding bis dizonium salts which were then coupled with 8-hydroxyquinoline to give the diether bis bidentate ligands. All products were characterized by FTIR, ^{1}H NMR, and elemental analyses. The detailed synthetic procedures will be described elsewhere. Few representative examples will be given here.

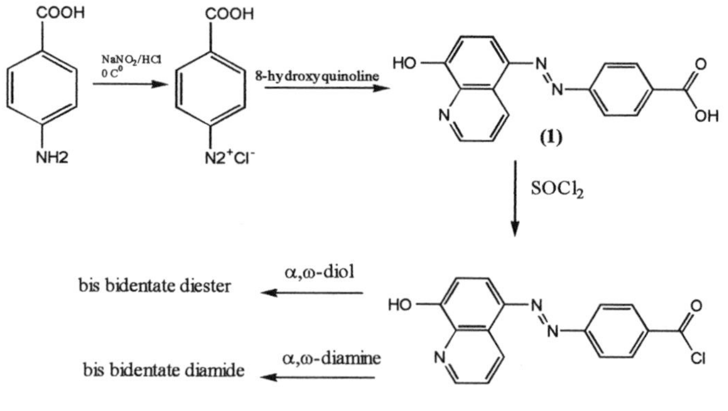

Scheme (I): Synthesis of Bis Bidentate Diesters and Diamides

Compound **1** was also used in the preparation of a bis bidentate diester with side chain chromophore as shown in Scheme (II). Polymers prepared using this compound could show some second order NLO properties. Since the metal usually retains partial positive charge, it remains to be seen if this would interfere with poling such polymers.

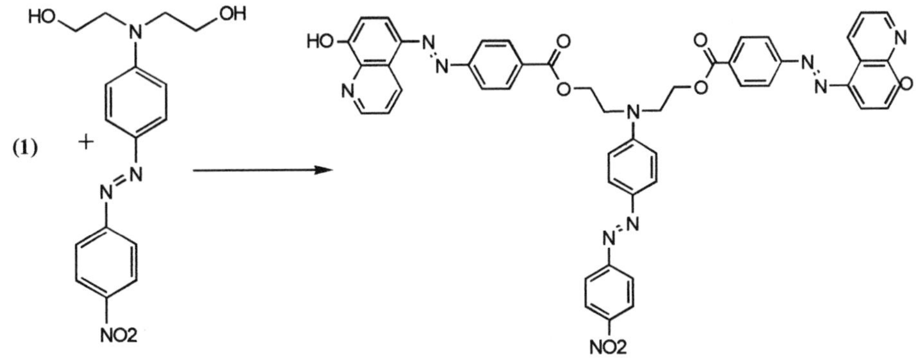

(1)

Scheme (II): Synthesis of a Bis Bidentate Ligand With a Side Chain Chromophore

Synthesis of the diester bis bidentate ligands:

The acid chloride of compound **1**, (4.00 mmol) and α,ω–alkane diol (2.00 mmol) were heated at reflux in a 3.0 mL pyridine for 30 min. The reaction mixture was then allowed to cool to room temperature. Upon adding 50.0 mL of chloroform, the reaction mixture was washed with 200 mL of 0.5M HCl followed by 200 mL of 0.5M NaOH. The organic layer was then separated and crystallization was effected by adding ethanol. The products were soluble in chloroform, and slightly soluble in ethanol. For the C6 spacer group: m.p. 212 (dec.), IR: 1700 cm^{-1}, 1270 cm^{-1} (ester). The absorption at 1790 cm^{-1} of the acid chloride completely disappeared. λ_{max}(ethanol)= 400nm, ^{1}H NMR(δ ppm in CDCl$_3$):3.65 d, 1.92 m, 7.45 d, 7.98 d, 7.20 d, 6.81 d and 8.10 doublet of doublets.

Solution Polymerization:

The hydrate acetate of the divalent metal dissolved in dimethyformamide (DMF), was slowly added to a stirred DMF solution containing the ligand and maintained at 120 C$^\circ$. In all cases precipitation occurred almost immediately upon mixing the two reactants. The reaction mixture was heated under reflux for an additional 30 min, then allowed to cool to room temperature overnight. The product was collected, extracted with either DMF or chloroform followed by ethanol (to remove any residual unreacted ligand) using Soxhlet apparatus. The solutions in the extraction apparatus were highly colored. The insoluble polymer was then dried in vacuum at 300 C$^\circ$.

RESULTS AND DISCUSSION

The polycoordination (polychelation) of metal centers with the bis-bidentate ligands is, from a synthetic point of view, relatively simple. It is surprising that this method is rarely employed in the preparation of metal containing polymers. Figure 2 shows the general structures of polymers targeted in this study. The metals used in this study are: Cu, Zn, Pb, Pd, and Ni. The spacer groups used were C5-C12.

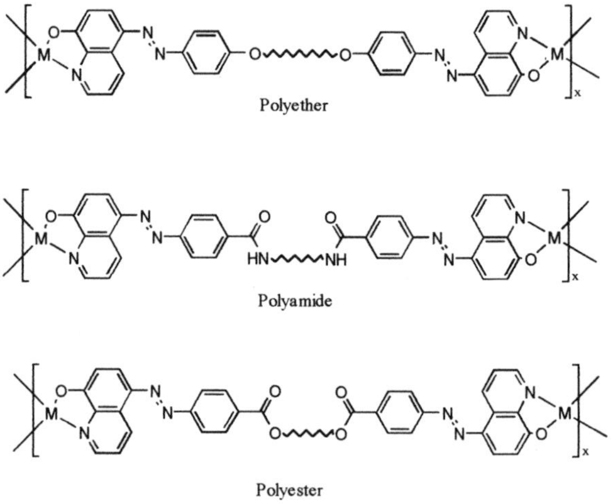

Polyether

Polyamide

Polyester

Figure 2: Structures of Metal Containing Polymers

The FTIR spectra of the DMF soluble fractions (during the Soxhlet extractions) and the insoluble solids were almost identical. This indicates that the lower molecular weight materials was extracted into DMF while the higher MW was not. GPC data (using polystyrene as standard) on the DMF soluble portion gave M_n values of 3000-9000, with the longer spacer groups giving the higher values[*]. We are currently addressing this problem of low molecular weight products and also improving the solubility of the insoluble fraction. One approach we are currently pursuing is the use of interfacial polymerization at the org./aq. Interface. The C-O stretching frequency was used as evidence for the formation of the metal containing polymers. For instance the polyester

---

[*] We have obtained dimers only in some runs, we believe that this could be a result of the formation of macrocyle in which two bis bidentate ligands form a ring like structure with two metal atoms at specific spacer lengths. We are currently looking further into that possibilty.

with 6 carbons in the spacer group, with Zn, the C-O stretching frequency was 537.1 cm^{-1}, which changed to 537.5 cm^{-1} upon using Pb and to 550 cm^{-1} with Pd. Based on elemental analyses, there was no evidence that the polymers suffered any decomposition at 300 C in vacuum. This is comparable to the decomposition temperatures for monomeric 8-hydroxyquinoline chelates [8]. If the same structure/ property relations hold true for this class of polymers as for other polymers, one can still greatly improve on this high thermal stability by using the linking group $-SO_2-$. Our results, thus far suggest the following: coordination polymerization provides a relatively simple route to metal containing polymers. The resulting polymers have excellent thermal stability, but rather low solubility. The solubility of these polymers in chloroform for example is greatly improved by removing the water of hydration (heating at 150 C or higher, as determined by weight loss measurements). The anhydrous polymers are more soluble in chloroform, but solutions slowly turn cloudy upon standing, presumably by regaining the water of hydration. To completely eliminate this problem our next step is to use trivalent metal centers.

CONCLUSIONS

We reported on our preliminary finding on a new group of metal containing polymers. These polymers have three major advantages: (1) ease of preparation, (2) versatile structural features, (3) excellent thermal stability. Two main obstacles remain to be addressed: one is the solubility problem and the second is obtaining higher molecular weight materials which are still soluble. The future directions of this research effort will focus on addressing these two issues.

ACKNOWLEDGMENTS

The financial support form the Commonwealth College at the Pennsylvania State University in the form of a Research Development Grant is highly appreciated. J.M. and L.A would like to thank the Director of Academic Affairs at PSU/Hazleton for providing partial travel support to attend the MRS meeting.

REFERENCES

[1] (a) C. E. Sheats, C. U. Pittman, M. Zeldin, B. Currell, Eds. *Inorganic and Metal-Containing Polymeric Materials*, Plenum, New York, 1990; (b) D. R. Kanis, P. G. Lacroix, M. A. Ratner, T. J. Marks, *J. Amer. Chem. Soc.* 116, p. 10089, 1994 and references therein.

[2] (a) E. Horowitz, T. P. Perros, *J. Inorg. Nucl. Chem.*, 26, p.139, 1964; (b) J. S. Oh, J. C. Bailar, Jr., *J. Inorg. Nucl. Chem.*, 24, p.1225, 1962; (c) V. V. Korshak, S. V. Vinogradova, T. M. Babchinister, Polym. Sci. (USSR) Engl. Transl., 2, p. 344, 1960.

[3] E. W. Berg, A. Alam, *Anal. Chim Acta*, 27, p. 454, 1962.

[4] G. L. Eichhorn, N. D. Marchand, *J. Amer. Chem. Soc.* 78, p. 2688, 1956.

[5] M. M. Bader, T. Hamada, A. Kakuta, *J. Amer. Chem. Soc.* **114**, p. 6475, 1992; M. M. Bader, *SPIE Proc.* **3473**, p.112, 1998

[6] (a) C. W. Tang, S. A. Van Slyke, *Appl. Phys. Lett.*, **51**, p. 913, 1987; (b) C. W. Tang, S. A. Van Slyke, *J. Appl. Phys.*, **65**, p. 3610, 1989; (c) C. Aachi, S. Tokito, M. Morikawa, T. Tsutsui, S. Saito, *Springer Proc. In Phys.*, **38**, p. 358, 1989.

[7] M. M. Bader, *Phosphorous, Sulfur and Silicon*, **116**, p.77, 1996.

[8] R. G. Charles, A. Langer, *J. Phys. Chem*, **63**, p. 603, 1959.

OPTICAL NONLINEAR PROPERTIES OF SELF-ASSEMBLED J-AGGREGATES CYANINE MOLECULES DOPED IN SPIN-CASTED SILICATE FILMS

I. HONMA, T. WATANABE*, H.S. ZHOU, A. MITO**, K. ASAI*, K. ISHIGURE**

Energy division, Electrotechnical laboratory, Umezono 1-1-4, Tsukuba, Ibaraki, 305-8568, Japan

E-mail: ihonma@etl.go.jp; Tel: +81-298-54-5797, Fax: +81-298-54-5805

*Department of Quantum Engineering and Systems Science, The University of Tokyo 7-3-1 Hongo, Bunkyo-ku, Tokyo, 113-0032, Japan

**National Research laboratory of Metrology, AIST, Umezono 1-1-4, Tsukuba, Ibaraki, 305-8568, Japan

ABSTRACT

J aggregates cyanine dyes have been successfully doped directly in sol-gel derived silicate thin films. The films are transparent, homogeneous and stable at room temperature. The J aggregates successfully incorporated in rigid SiO_2 framework were characterized by absorption spectroscopy and fluorescence microscopy. The films are expected to be useful for non-linear optical devices, multiple photon recording and photoelectric cell.

INTRODUCTION

Cyanine dyes have played an important role as a photosensiter in photography[1] and in lithography[2]. Recently, the dyes have drawn much attention for non-linear optical devices due to large third order nonlinear optical susceptibility[3,4], and also for photoelectric cells[5,6], multiple photo-recording devices through photochemical hole burning[7]. The J aggregate materials for above applications have to satisfy the following requirements: (1) thin films due to handling, (2) high concentration in any matrix, (3) briefly preparation and (4) high stability of J aggregates. It is well known the morphologies of the aggregates can be controlled in solutions[8], LB films[9], crystals[10] and dispersed polymer films[11,12]. However, the process optimization was still needed further improvements.

In the present experiment, J aggregates of a cyanine dye have been successfully doped in rigid inorganic framework of spin casted silicate films. The aggregate structure of the molecules and optical absorption of the doped films were characterized.

EXPERIMENT

The two different cyanine molecules of dye 1: 1-methyl-1'-octadecyl-2, 2'-quinocyanine perchlorate and dye 2: 1, 1'-dimethyl-2, 2'-quinocyanine bromide dyes were used in the present experiments, which was shown in fig. 1. The difference of dye 1 and dye 2 is that the former has longer alkyl chains and larger anions. These dyes were used without farther purification. The dye 1 is frequently used in J aggregates in LB films[13]. So we start with dye 1 of longer alkyl chain for doping aggregate molecules in rigid silicate framework.

The preparation method of doping J-aggregate in silica was simply summarized in fig. 2.

For instance, 0.25 ml tetraethyl orthosilicate (TEOS) was mixed with 5 ml ethanol (EtOH) and stirred for 10min. And, the solution was added with 0.0725 ml HCl (0.14 N) and stirred for 60 min. Then, this solution becomes the sol mother-solution. It was added with various amount of the dye (dye 1 or dye 2) and stirred for another 180 min. Finally, the sol-solution was used for film deposition on grass substrates ($35 \times 25 \times 1.5$ mm^3) by spin casting with a low-speed of 250-500 rpm for the initial 3 sec and at a high-speed of 500-4000 rpm for the final 10 sec. All of the procedures were carried out at room temperature (28°C) and was schematically drawn in fig. 3.

The liquid films of the sol-solutions and the silica films were characterized with UV-visible absorption spectroscopy (Hitachi U-4000). Since the sol-solutions are too concentrated to measure its absorbance, the liquid films formed from the sol-solutions were measured. The absorbance was normalized by the extinction of peak height assigned to the quinoline ring band at 330 nm ($\varepsilon_{dye\ 1} = 1.1 \times 10^4$, $\varepsilon_{dye\ 2} = 1.0 \times 10^4$). Since no band is superimposed on quinoline ring band in any aggregation for these dyes, this evaluation seems reasonable. The thickness of films was estimated with ellipsometry (Rudolph research Auto EL-III). The dye density was evaluated from these results. Moreover, the morphology of the dye silica films was observed by fluorescence microscopy.

Furthermore, third order susceptibility $\chi^{(3)}$ of the films was measured by z-scan method[14] using 180 femtoseconds pulse laser beam. The measurement was performed at 77 K in vacuum environment ($<10^{-5}$ Torr). In the Z-scan measurement, we used fused silica as a standard material. Its $\chi^{(3)}$ value was evaluated as 1.3×10^{-14} esu from fitting the data of refs[15],[16].

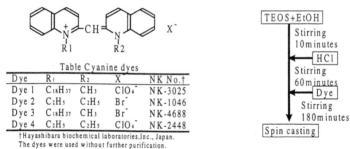

Table Cyanine dyes

Dye	R_1	R_2	X	NK No.†
Dye 1	$C_{18}H_{37}$	CH_3	ClO_4^-	NK-3025
Dye 2	C_2H_5	C_2H_5	Br^-	NK-1046
Dye 3	$C_{18}H_{37}$	CH_3	Br^-	NK-4688
Dye 4	C_2H_5	C_2H_5	ClO_4^-	NK-2448

†Hayashibara biochemical laboratories,Inc., Japan.
The dyes were used without further purification.

Fig. 1 Cyanine dyes used in this work.

Fig.2 Film fabrication by spin casting.

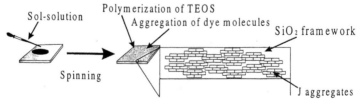

Fig.3 The fabrication method of J aggregate doped silica thin film.

RESULTS and DISCUSSION

In both of dye 1 films and dye 2 films, the sol-solutions were red; the dye doped silica films were purple. The color transition indicating the J aggregation in the silica films was observed during the fast spinning of the spin casting procedure in the film fabrication. Naturally, uniformity and transparence of the films depended on the spinning speed during the fast spinning procedure, also. While the films fabricated at a low speed below 1500-rpm possess fine cracks, thickness-roughness and opacity, the films fabricated at a speed of 1500 rpm or above were clear and homogeneous in thickness, and didn't form such cracks. The J aggregates in the films doped with condensed dyes are stable at room temperature.

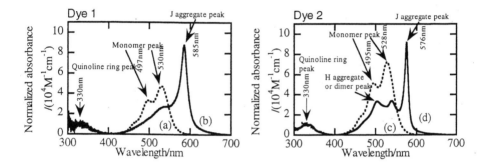

Figure 4 shows the absorption spectra of sol-solutions (liquid films) and silica films. It indicates that the dye molecules don't form any aggregates in the sol-solutions, whereas they form the J aggregates in the silica films. The line (a) and (c) show the spectra for a dye 1 sol-solution and a dye 2 sol-solution, respectively. Both of them, which are similar to each other, appear monomeric: the band at 330nm is assigned to the quinoline rings in the dye molecule structures. The bands at around 495nm and at around 530nm are to the π-electron vibration between the N atoms in these molecule structures in monomeric condition. Since dye 1 molecule possesses the same chromophore as dye 2 molecule, these dye monomers indicates consequently the identical spectra in the range from 300 to 700 nm.

However, difference in the alkyl chains and counter anions between the molecule structures of these dyes cause difference of the aggregation conditions and in the solubility's. The line (b) shows the spectrum of a silica film fabricated from the dye 1 solution. The sharp J band (HWHH=27cm^{-1}) stands at 585 nm and the two shoulders at the monomer bands. Because of overlap of monomeric band and J aggregate band, the monomers and the J aggregates appear to coexist in the films. On the other hand, the line (d) shows the spectrum of the silica film fabricated from the dye 2 solution. The prominent J band of dye 2 stands at 576nm. In addition to the J band and the two monomeric bands, either H aggregates or dimer band appears to overlap on these bands at around 505 nm[17]. Thus, the monomers and the various aggregates would coexist in the films. The J band of dye 2 (HWHH=14cm^{-1}) is considerably narrower than that of dye 1. J aggregates formed by dye 2 molecules exist in more homogeneous condition than that by dyes 1 molecules.

Moreover, dye density in the silica films was evaluated from the spectra. It increased with increasing dye concentration in the sol solutions used for film fabrication. Then, the maximum dye densities in the silica films were determined by the dye solubilities in the sol-solutions. The solubility of dye 1 in the sol mother-solution at room temperature (28°C) is a weight ratio (dye 1/sol-solution) of 0.16%; that of dye 2 is 0.30%. In our experiment, the maximum dye density in the dye 1 films was 0.1 nmol/cm^3 corresponding to the saturated dye 1 solution. On the other hand, that in the dye 2 films was 0.8 nmol/cm^3 corresponding to the saturated dye 2 solution. In addition to deference between the solubilities of these dyes, since dye 1 molecule are more bulky than dye 2 molecule, the density of the dye 2 in the silica films would be higher than that of the dye 1.

Figure 5 (a) and (b) show that the absorbance of J band strongly depends on both dye concentrations and spinning speed. Fig. 5 (a) shows that, in the dye 1 (Plot (a)), the absorbance (585nm) associated with J aggregates linearly raises with increasing of the dye 1 concentration in low concentration range at a weight ratio (dye 1/sol-solution) from 0.000% to 0.005% and saturates at a weight ratio of 0.16%. On the other hand, plot (b) shows the dependence of J aggregate absorbance (576 nm) on dye 2 concentration in the sol-solutions; the growth of J aggregates prominently sets at 0.10% and saturates at 0.30%. From comparison between (a) and (b), it is found that the dye 1 molecules prefer to aggregate rather than the dye 2 molecules. However, the J aggregates in dye 2 are more stable than in dye 1. This is because the dye molecules would be protonated by HCl added into the sol-solutions for promotion of condensation polymerization of TEOS, so that the dye molecules in rare dye condition would be protonated farther than in condensed dye condition.

Figs. 5 (b) shows that the J aggregation tends to be promoted by increasing of the spinning speed for the films fabricated from a sol-solution of a fixed concentration in both of dye 1 films and dye 2 films. Plot (a) and plot (b) show the spinning speed dependence of J aggregation in dye 1 films in dye 2 films, respectively. The J aggregation drastically increases with increasing of spinning speed below 1000 rpm, and gradually above 1000 rpm. However, the film thickness unfortunately decreases with increasing of the spinning speed instead of efficient J aggregation in both of the dye films, so that the dye surface-density is subsequently sacrificed.

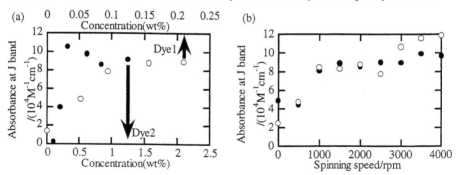

Fig. 5. J band absorbance dependence of dye concentration and spinning speed. (a) Absorbance as a function of concentration, (b) absorbance as a function of spinning speed. ○: absorbance of dye 1 doped silica films at 585nm; ●: absorbance of dye 2 doped silica films at 576nm.

The process of silica film formation was as follows; (1) the solution was dropped on the substrate (2) the substrate rotates and the excess amount of the solution was removed. The remained solution forms a liquid film on the substrate. Its thickness is defined by the spinning speed. (3) By evaporate of EtOH from the liquid film, the solution is rapidly condensed. And then, both of the aggregation of dye molecules and the condensation polymerization of TEOS would compete with EtOH evaporation. (4) Finally, the SiO_2 framework on the substrate would be consequently inlaid with dye molecules.

There would be three principal reactions competing in pathway from a sol-solution to a silica film; (1) evaporation of EtOH, (2) aggregation of dye molecules and (3) condensation polymerization of TEOS. This competition depends on reaction factors such as dye concentration in sol-solution, thickness of the liquid film on the substrate, temperature and pH. These parameters affect consequently the dye aggregate condition in SiO_2 framework. As show in fig. 2, when a low concentration sol-solution is used in film fabrication, J aggregation is difficult in silica films. This is because the condensation polymerization of TEOS would become more dominant than dye aggregation, so that the isolated dye molecules would be confined in the SiO_2 framework. Furthermore, spinning speed dependence of J aggregation can be interpreted in the way following. Since a thin liquid on the substrate dries further than a thick film, EtOH in thin liquid films seems to evaporate further than in thick films. Fast spinning of the substrate decreases film thickness of the liquid film on the substrate. As a result, the rate of aggregation of dye would be much faster than that polymerizes of TEOS. Thereby, aggregation would be promoted by fast spinning.

In our fabrication, choice of solvent is quite important. Addition to affinity for water and alkoxysilane such as TEOS, solvents with high solubility with dyes have to be employed. This is because, when a solvent with low solubility is used in film fabrication, it is difficult for dye molecules to aggregate, or even if J aggregation is possible, the aggregates are quite unstable. Furthermore, the counter anion species with dye molecules are significant factor in J aggregation. In preliminary experiment, J aggregates in silica films was prepared using sol-solutions of 1-methyl-1'-octadecyl-2, 2'-quinocyanine perchlorate (dye 3) and 1, 1'-dimetyl-2, 2'-quinocyanine bromide (dye 4) with the same fabrication procedure as dye 1 and dye 2. However, they don't form aggregates in the silica films. Because of quite low solubility of dye 3 in mother-solution, dye 3 molecules didn't form any aggregates into the silica films although dye 4 solves farther in sol mother solution than dye 3, the H aggregates or the dimers were formed in the silica films instead of J aggregates.

Finally, figure 6 shows wavelength dependence of absolute value of third order susceptibility $|\chi^{(3)}|$. The sample is the dye 2 doped silica film. Its $|\chi^{(3)}|$ is 5×10^{-7} esu near on-resonant wavelength of 577 nm. The nonlinearity is mainly due to nonlinear absorption arising from Frenkel exciton generation. The films show bleached absorption at the longer and induced absorption at the shorter wavelength side in the J band. That suggests bi-exciton formation by two-photon absorption.

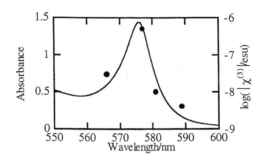

Fig. 6. Wavelength dependence of $\chi^{(3)}$. The sample is the dye 2 doped silica film, which is formed on a quartz substrate at spinning speed of 2000 rpm. Its film thickness is 170 nm. Solid line is absorbance; ● is $|\chi^{(3)}|$.

CONCLUSIONS

The thin silica films doped with condensed J aggregates of cyanine dyes are synthesized for the first. In spite of containing plenty of J aggregate, these films are transparent and homogeneous in thickness. In addition, the synthesis of the film is brief and simple. Thus, the films are expected to be useful as materials for many applications.

REFERENCES

1. D. M. Sturmer, D.W. Heseltine, in: T. H. James (Ed.), Theory of Photographic Process, 4th Ed., Macmillan, London, 1977.
2. U. C. Fischer and H. P. Zingsheim, J. Vac. Sci. Biophys. Chem. 4, p.881(1981).
3. F. C. Spano and S. Mukamel, Phys. Rev. A40, p.5783 (1989).
4. R. Gagel, R. Gadonas and A. Laubereau, Chem. Phys. Lett. 217, p228(1994).
5. D. L. Morel, E. L. Stogryn, A. K. Ghosh, T. Feng, P. E. Purwin, R. F. Shaw C. Fishmann, G. R. Bird and A. P. Piechowski, J. Phys. Chem., p.923(1984).
6. A. P. Piechowski, G. R. Bird, D. L. Morel and E. L. Stogryn, J. Phys. Chem., p.934(1984).
7. S. De Boer and K. J. Vink and D. A Wiersma, Chem. Phys. Lett. 137, p.99(1987).
8. B. Kopainsky, J. K. Hallermeir and W. Kaiser, Chem. Phys. Lett. 87, p.7(1982).
9. J. Terpstra, H. Fidder and D. A. Wiersma, Chem. Phys. Lett. 179, p.349(1991).
10. A. P. Marchetti, C. D. Salzberg and E. I. P. Waoler, J. Chem. Phys. 64, p.4693(1976).
11. K. Misawa, H. Ono, K. Minoshima and T. Kobayashi, Appl. Phys. Lett. 63, p.577(1993).
12. S. De Boer and D. A. Wiersma, Chem. Phys. 131, p.135(1989).
13. K. Asai, T. Watanabe, D. Heroic, K. Ishigure, Thin Solid Films 283, p.124(1996).
14 M. Sheik-Bahae, A. A. Said, T. Wei, D. J. Hagan and E. W. Van Stryland, IEEE. J. Quantum Electrons, 26, p.760(1990).
15 A. Mito, K. Hagimoto, and C. Takahashi, Nonlinear Opt., 13, p.3(1995).
16 D. Milam, Appl. Opt., 37, p.546(1998).
17 . B. Kopainsky, J. K. Hallermeir and W. Kaiser, Chem. Phys. Lett. 83, p.498(1981).

ULTRAFAST OPTICAL RESPONSES OF THE SQUARYLIUM DYE J-AGGREGATES FILMS

S.Tatsuura*, M.Furuki**, M.Tian*, Y.Sato*, L.S.Pu*
* Corporate Res. Labs., Fuji Xerox Co.,Ltd., 430 Sakai, Nakai-machi, Kanagawa 259-0157, Japan, tatsuura@rfl.crl.fujixerox.co.jp
** FESTA Laboratories, 5-5 Tokodai, Tsukuba 300-2635, Japan

ABSTRACT

Ultrafast optical responses of the organic films containing squarylium dye J-aggregates prepared by simple spincoating method were investigated. Some SQ films exhibited a sharp absorption peak originated from J-aggregates and especially the SQ derivative which has dibutylamino bases at the end of benzene rings forms good J-aggregates in the films without any extra treatment. Time-resolved difference absorption spectra were measured about the SQ44 film. Clear bleached absorption peak and its fast recovery were observed. The decay time was determined to be about 220 fs at 50 MW/cm^2 excitation intensity. Sample preparations and all the measurements were performed in air at room temperature.

INTRODUCTION

Essential factors for the nonlinear optical materials in the field of all-optical device applications should be a high speed response and a large $\chi^{(2)}$ or $\chi^{(3)}$ coefficient. In the case of semiconductor materials, many efforts have carried out mainly for the realization of the fast optical responses. But recently ultrafast optical responses were observed about some kinds of J-aggregates of organic molecules. Pseudoisocyanine bromide (PICBr) was reported to form J-aggregates in ethyleneglycol/water glasses at low temperature. Their optical properties were investigated and the fast decay component of the bleached and induced absorption was determined to be about 1.3 – 1.5 ps at 77 K[1-5]. Last year squarylium (SQ) dye J-aggregates were studied by our group as an another candidate[6-8]. The SQ derivatives with phenyl ring and alkyl side chains were synthesized and the formation of excellent J-aggregates was observed at room temperature in a mono-molecular layer at the air-water interface, so called a Langmuir-film. The decay times of the bleached and induced absorption of the SQ J-aggregates were calculated to be 280 fs and 200 fs respectively, and the percentage of the absorbance change toward the original absorption reached up to 30 %. But for the practical applications, Langmuir-films have some difficulties about the stability and the processability. PICBr J-aggregates mentioned above was an solid bulk but it is fluent at room temperature.

In this report, the preparation of a solid J-aggregates film utilizing simple spincoating method and the optical dynamics of the film are studied. The results indicates that the SQ dye J-aggregates is the attractive material for an ultrafast optical application because of their performance and of their cost-merit.

EXPERIMENT

The chemical structure of the SQ dye derivatives used in the experiments is shown in Figure 1. The synthesis of bis[4-(N-alkylamino)phenyl] squarylium was described in another paper[9]. SQ film was spincoated onto a glass or a sapphire plate at 3000-4000 rpm from a 1,2-

dichloroethane solution of 1-2 wt%. Optical absorbance of a typical film was 1.5-2.5. SQ films were heated in alkali or acid vapor if needed on purpose to assist forming J-aggregates. UV and fluorescent spectra were used to confirm the formation of J-aggregates.

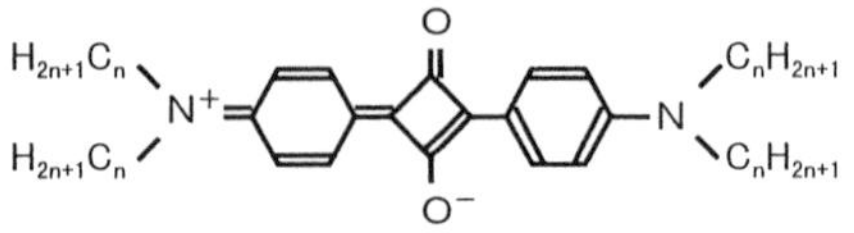

$(n = 3,4,7)$

Figure 1. The chemical structure of the SQ dye

The optical dynamics of femtosecond time scale was examined by the femtosecond pump-probe measurement. The excitation (pump) pulse was generated and tuned by the optical parametric amplifier pumped by the output pulses from the T:Sapphire regenerative amplifier. Center wavelength of the excitation pulse was 770 nm and cross-correlation width was 200fs under 1kHz repetition rate. White light continuum from self-phase modulation was produced by focusing the amplified pulse into a sapphire plate. Typical energy of the excitation pulse was 2 μ J and the excitation power density was changed between 15 MW/cm^2 and 220 MW/cm^2. For reducing the optical noises of self-diffraction and fluorescence by the excitation pulse, the background signal was measured separately for each spectrum. Sample preparations and all the optical measurements were performed in air at room temperature.

RESULTS

The SQ derivatives used in the experiments had almost identical molecular structures. They were differ from each other only in the substituted alkyl side chain length. But remarkable differences were observed in the absorption spectra. Figure 2(a) shows the absorption spectra of the spincoated films of the SQ derivatives with dipropylamino bases (SQ33), dibutylamino bases (SQ44) and diheptylamino bases (SQ77), together with the spectrum of SQ33 1,2-dichloroethane solution. SQ33, SQ44 and SQ77 show almost the same absorption in solution because the difference of alkyl chain length makes only a slight influence to the electron π – conjugation of the molecule. Two features could be pointed out about the spectra of Figure 2(a). One is the peak shift of the film compared to the solution. In the case of SQ33, the spectrum is red shifted about 70 nm from the solution to the film. It means that the initial formation of aggregates occurs in the film. The formation of aggregates was observed so far only in the two-dimensional Langmuir films[10]. But Figure 2(a) indicates that the aggregation progresses in the three-dimensional spincoated films too. Second feature is, of course, the differences of the spectrum between the SQ derivatives. SQ33 has one broad absorption peak around 710 nm, which is the result from the initial aggregation. On the other hand, SQ44 shows two peaks, one is the sharp peak of 770 nm and second is the broad one around 650 nm. To identify these peaks, the fluorescent spectrum of SQ44 film was measured (Figure 2(b)). When excited at 650 nm, fluorescence was detected at the peak wavelength of 777 nm. Quantum efficiency increased rapidly as excitation wavelength close to 770 nm. From the pretty small stokes shift of 770 nm absorption, this absorption is attributed to the SQ J-band. The absorption at the higher energy side should be come from the residual dimers or trimers. When excited at these non-J-band, efficient energy transfer from non-J-band to J-band could be occurred. In Figure 2(a), SQ derivatives with longest alkyl side chain, SQ77, forms (not J-)aggregates too but exhibits broader absorption peak than SQ33. The most suitable side chain length for forming J-aggregates might be exist. In fact, no SQ derivatives was found so far except SQ44, which forms J-aggregates in a spincoated film as it is.

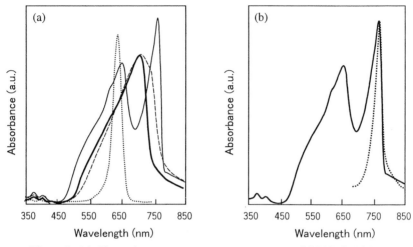

Figure 2. (a) Absorption spectra at room temperature of SQ33 (bold line), SQ44 (solid line), SQ77 (broken line) spincoated films and SQ33 1,2-dichloroethane solution (doted line), (b) Absorption (bold line) and fluorescent (doted line) spectra of SQ44 spincoated film with the excitation wavelength of 650 nm.

For some SQ derivatives, the formation of the aggregates was encouraged by the existence of heat and/or acid vapor. The influences of these extra treatments were varied according to the molecular structure, namely, alkyl side chain length. SQ33 spincoated film exhibits the narrower absorption peak of J-band than SQ44 by heating in acid vapor, but SQ44 and SQ77 spectra do not change under the same treatment. Molecular morphology and kinetics of the SQ derivatives in a film should be governed mostly by the alkyl chain length but details are discussed elsewhere.

Once SQ J-aggregates formed in the spincoated films, they are rather stable. With six months passed after the spincoating, only a small absorption drop was observed about SQ44 J-aggregates when the films were kept in a dark space at room temperature. In the case of the spincoated films, SQ J-aggregates should be formed inside the films, so they could avoid the oxygen attack unlike the case of the Langmuir-films. It might contribute to the stability of the SQ J-aggregates.

Time-resolved difference absorption spectra were measured about the SQ44 J-aggregates film. Figure 3 shows the typical difference absorption spectra. The peak wavelength of J-band of the sample was 766 nm, therefore the excitation wavelength, 780 nm, lies lower energy side of J-band. Clear bleached absorption is observed with the peak at 775 nm and it almost diminishes within 1 ps after the excitation. The induced absorption, which is the transition from the 1 exciton states to the higher exciton states, can not be seen clearly in Figure 3 because the J-band of SQ44 has the gentle slope at higher energy side unlike the sharp absorption edge at lower energy side. To calculate the decay time of the bleached absorption, the absorbance change was plotted as the function of the delay time and curve fitted by the model functions (Figure 4). As a result, 360 fs was obtained as the fast decay component and 12 ps as the slow one. This fast component was attributed to the cooperative emission (superradiance) and the slow one to be the spontaneous radiation. This indicates that the ultrafast optical processes, which is observed in the two-dimensional Langmuir-films, is occurred in the simple three-dimensional spincoated films too.

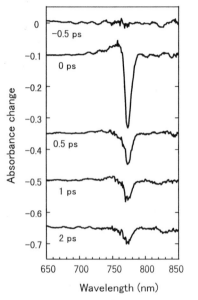 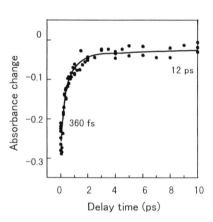

Figure 3. The difference absorption spectra of the SQ44 spincoated film excited at 220 MW/cm² at room temperature.

Figure 4. The absorbance change as the function of the delay time.

The fast decay component of the bleached absorption is plotted as the function of the excitation intensity in Figure 5. The decay time increases as the excitation intensity is enlarged. The fastest decay time, 220 fs, was obtained with the excitation intensity of 50 MW/cm². Because the time resolution of the optical apparatus used in the pump-probe measurement was about 200 fs, the decay time faster than 220 fs could not be measurable. In the excitation range shown in Figure 5, the saturation of the absorption change could not be observed. In the case of the PICBr J-aggregates, the decrease of the decay time was observed along with the increase of the excitation intensity, which was explained for the exciton-exciton annihilation by hopping processes [3]. But at least more than 1 ps is needed for the exciton hopping, this process does not dominant in the SQ J-aggregates. The dependency shown in Figure 5 is explained for the extra decay from the higher exciton state, namely, at higher excitation intensity the decay time from the $n(\geqq 2)$-exciton state to the 1-exciton state should be added to the decay time. The decay time of the $n(\geqq 2)$-exciton state was determined to be about 200 fs for PICBr J-aggregates[4]. This value is close to the increased amount of the decay time when the excitation intensity is increased for the SQ J-aggregates. Therefore, in the case of the SQ derivatives, the exciton relaxation process should be localized in a single J-aggregates.

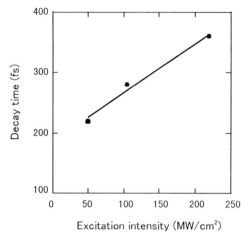

Figure 5. The change of the decay time of the bleached absorption. Solid line is just a guide to eye.

As the material for ultrafast optical devices, the SQ J-aggregates in this report has some noticeable features. First, the films which have sub-picosecond optical responses are easily prepared by a simple spincoating method. Second, they can work in air at room temperature and no special environments are needed for preparing and keeping the films. At last, the large size film is readily earned, which is suitable for making the one or two dimensional optical switch array.

CONCLUSIONS

The preparations and the optical responses of the SQ J-aggregates spincoated films were investigated. The SQ derivatives containing dibutylamino alkyl chains exhibited a sharp absorption peak of J-aggregates. It was clarified that the morphology and kinetics of SQ derivatives in a film were governed mostly by the substituted alkyl chain length. Optical response of the SQ44 film was investigated by time-resolved difference absorption spectra. Clear bleached absorption was observed and the decay time was calculated to be about 220 fs at 50 MW/cm^2 excitation intensity. Film preparations and all the measurements were performed in air at room temperature. The SQ J-aggregates spincoated film has some noticeable features compared with semiconductor materials and it is an attractive candidates for the material of ultrafast optical applications.

ACKNOWLEDGMENTS

This work was supported by the New and Industrial Technology Development Organization in the framework of Femtosecond Technology Project.

REFERENCES

1. H.Fidder, J.Knoester, and D.A.Wiersma, J.Chem.Phys. **98**, p.6564 (1998)
2. S.Kobayashi and F.Sasaki, J.Lumin. **58**, p.113 (1993)
3. S.Kobayashi and F.Sasaki, J.Lumin. **60&61**, p.824 (1994)
4. K.Minoshima, M.Taiji, K.Misawa, and T.Kobayashi, Chem.Phys.Letters **218**, p.67 (1994)
5. K.Minoshima, M.Taiji, A.Ueki, K.Miyano, and T.Kobayashi, Nonlinear Optics **14**, p.39 (1995)
6. M.Furuki, L.S.Pu, F.Sasaki, S.Kobayashi, and T.Tani, Mat.Res.Soc.Symp.Proc. **488**, p.777 (1998)
7. M.Furuki, L.S.Pu, F.Sasaki, S.Kobayashi, and T.Tani, Appl.Phys.Letters **72**, p.2648 (1998)
8. M.Furuki, L.S.Pu, F.Sasaki, S.Kobayashi, and T.Tani, Mol.Cryst.Liq.Cryst. **316**, p.67 (1998)
9. H.E.Sprenger and W.Ziegenbein, Angew.Chem.Int.Ed.Engl. **5**, p.894 (1966)
10. M.Furuki, S.Kim, L.S.Pu, H.Nakahara, and K.Fukuda, J.Chem.Soc.Japan Chem.Ind.Chem. **10**, p.1121 (1990)

ELECTROSTATIC SELF-ASSEMBLY OF MULTILAYER NONCENTROSYMMETRIC THIN FILMS AND DEVICES

K. M. LENAHAN[a], T. ZENG[a], Y. LIU[b], Y.-X. WANG[a], W.ZHAO[a], R.O.CLAUS[a]
[a] Fiber & Electro-Optic Research Center, Virginia Polytechnic Institute & State University, Blacksburg, Virginia. [b] NanoSonic, Inc., P.O. Box 618, Christiansburg, Virginia.

ABSTRACT

The electrostatic self-assembly of multilayer thin films by alternate adsorption from polyelectrolyte solutions spontaneously leads to the formation of noncentrosymmetric structures if the molecules themselves have net dipole moments. Significant second-order nonlinear optical susceptibility has been observed in such films, using both commercially available chromophores and molecules specifically designed to yield an enhanced net dipole moment. Recent results indicate the capability to fabricate piezoelectric films using the same method. The nature of the deposition process results in an alignment of the chromophores that is stable over time and to temperatures up to 150°C, in contrast with poled polymers. ESA films offer the additional major advantages of excellent homogeneity and low optical loss, high thermal and chemical stability, and low cost.

INTRODUCTION

ESA processing inherently yields noncentrosymmetric molecular structures that possess remarkably large $\chi^{(2)}$ second order NLO response, without the need for electric field poling. Such structures have been shown to exhibit inherent long-term stability, in contrast to conventional poled polymers. In recent years, novel piezoelectric materials have attracted much interest for poly-MEMs devices and actuators [1,2]. Design of new smart materials and devices requires a flexible manufacturing process and film thickness on the nanometer scale. ESA has been used successfully to synthesize a number of other multifunctional high-performance materials [3-7]. This low-cost process presents significant advantages over conventional thin film synthesis techniques, including excellent molecular-level uniformity, nanoscale thickness control, and design flexibility.

Figure 1 shows the basic concept behind the recently developed ESA process [8-10]. On the left, the substrate surface has been thoroughly cleaned and charged through chemical processing. The negatively charged substrate is then dipped into a solution containing polycations that are attracted to the anion surface and self-assemble into a monolayer. After thorough rinsing in ultrapure water to remove any loose molecules not ionically bonded to the surface, the substrate is dipped in a polyanionic solution. Subsequent polycation and polyanion monolayers are added in bilayer pairs by alternately dipping the substrate into polyanion and polycation solutions, to produce a multilayer thin-film structure as shown. Although Figure 1 suggests that each of the successive layers is a long polymer-like molecular chain, individual nanoparticles may be incorporated into any or all of the monolayers, allowing wide design opportunities for thin films with specific or multifunctional properties.

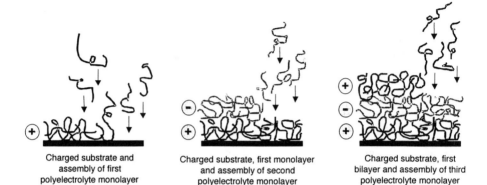

| Charged substrate and assembly of first polyelectrolyte monolayer | Charged substrate, first monolayer and assembly of second polyelectrolyte monolayer | Charged substrate, first bilayer and assembly of third polyelectrolyte monolayer |

Figure 1. Basic ESA schematic for buildup of multilayer assemblies by consecutive adsorption of anionic and cationic nanoparticle-based polyelectrolytes.

The following sections describe the synthesis and characterization of novel noncentrosymmetric ESA thin films. Initial experiments were performed using commercially available NLO polymer dyes. The work was then expanded to incorporate molecules with potential for piezoelectric applications and organic molecules specifically designed and synthesized for ESA NLO applications.

NONLINEAR OPTICAL ESA THIN FILMS

ESA NLO films were synthesized on glass substrates using commercially available polymer dyes (Poly S-119, Poly R-478 and PCBS, available from Sigma-Aldrich) and novel polymers synthesized during this study, called "Polydyes," with passive (no $\chi^{(2)}$ response) counterion layers. The films were characterized structurally and optically to verify transmission characteristics, uniform growth, and film thicknesses. UV/vis spectroscopy is a fast and simple method to verify the amount of chromophore deposited, and is a good indication of consistent chromophore density through he film thickness. Figures 2 and 3 show the linear growth of both optical absorbance and film thickness of representative samples, verifying that each bilayer is an identical, reproducible building block. In addition, the maximum differences between data collected at different locations across each film is at most a few percent, indicating excellent uniformity of the films (Fig. 4). This was confirmed by multiple AFM surface morphology visualizations, as illustrated in Figure 5. The ESA method therefore produces novel NLO thin films that are homogeneous through the film thickness as well as across the width and length of the film.

Both the commercial and newly synthesized polymer films were characterized using the same SHG setup described previously [11,12], with a fundamental wavelength of 1200 nm. The results are summarized in Table 1. The ESA process is shown to be compatible with numerous NLO polymer dyes, producing homogeneous films with exceptionally stable $\chi^{(2)}$ properties [13].

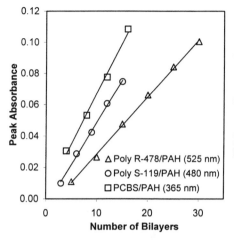

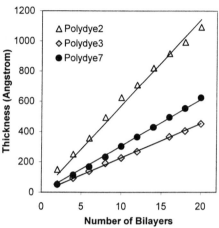

Figure 2. UV/vis absorbance increase during ESA NLO polymer film growth. Cationic monolayers consist of polyallylamine hydrochloride (PAH).

Figure 3. Film thickness increase during ESA NLO polymer film growth. Cationic monolayers consist of polydiallydimethyl - ammonium chloride (PDDA).

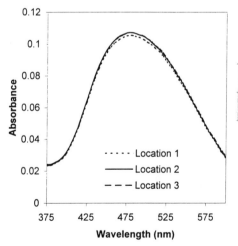

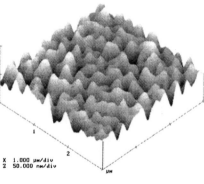

Figure 4. UV/vis absorbance at three separate locations on Polydye1 ESA film.

Figure 5. AFM image of 18 bilayer Polydye1 ESA film. X: 1 μm/div, Z: 50 nm/div.

Table 1. ESA NLO thin film comparison.

Sample	Number of bilayers	Film thickness	$\chi^{(2)}$ (pm/V)
Poly S-119	68	88 nm	**0.56**
PCBS	50	39 nm	**0.9**
Poly R-478	70	63 nm	**0.53**
Polydye1 (A)	15	43 nm	**3.6**
Polydye1 (B)	15	35 nm	**1.7**
Polydye2	52	296 nm	**0.34**
Polydye3	30	69 nm	**0.48**

PIEZOELECTRIC AND ELECTROSTRICTIVE ESA THIN FILMS

ESA processing has also been investigated for other noncentrosymmetric material applications, using polymer materials known for their piezoelectric and electrostrictive properties, as well as those newly investigated as part of this study. Material choices were evaluated using thick, spin-coated films and the normal load method for measuring the piezoelectric coefficient [14]. Polymer/polymer and polymer/inorganic nanocluster ESA films less than 100 nm thick were fabricated with the selected materials. For these ultrathin films, optical interferometry is required for measuring the piezoelectric coefficients [15,16]. Preliminary results indicate that unpoled organic/inorganic ESA composites produce positive piezoelectric results when measured with an interferometer system calibrated using PZT.

CONCLUSIONS

Both commercially available and newly synthesized NLO polymer dyes, as well as potential piezoelectric materials, have been incorporated into ESA films and their properties demonstrated using SHG and optical interferometry. This low-cost method produces thermodynamically stable, extremely homogeneous thin films with multifunctional characteristics determined by material selection and easily controlled deposition parameters.

ACKNOWLEDGEMENTS

The authors would like to thank Dr. J.R. Heflin and his students for their work on the SHG measurements on these films. This work was supported in part by ARO Grant DAAG55-92-1-0101, AFOSR contract F49620-97-C-0047, NSF contract DMI9860084.

REFERENCES

1. D.J. Jones, S.E. Prasad and J.B. Wallance, *Key Engineering Materials* **122-124**, 70-145 (1996).
2. P. Muralt, *Integrated Ferroelectrics* **17**, 297 (1997).
3. Y. Liu, A. Wang, R. Claus, *J. Phys. Chem B* **101**, 1385 (1997).
4. Y. Liu, Y.X. Wnag, R.O. Claus, *Chem. Phys. Lett.* **298**, 315 (1998).
5. A. Rosidian, Y. Liu and R.O. Claus, *Adv. Mater.* **10**, 1087 (1998).

6. K.L. Cooper, Ph.D. dissertation, Virginia Tech, April 1999.
7. Y. Liu and R.O. Claus, *Proc. SPIE* **3675**, March 1999.
8. R. K. Iler, J. Colloid Interface Science 1966, 21, 569.
9. G. Decher and J. Schmitt, Thin Solid Films 1992, 210/211, 813.
10. Y. Liu, Ph.D. dissertation, Virginia Tech, May 1996.
11. J. R. Heflin, Y. Liu, C. Figura, D. Marciu and R.O. Claus, in *Organic Thin Films for Photonics Applications*, Vol. 14, OSA Technical Digest Series 1997, 78-80.
12. J. R. Heflin, Y. Liu, C. Figura, D. Marciu and R.O. Claus, *Proc. SPIE* **3147**, 10 (1997).
13. C. Figura, D. Marciu, Y. Liu, Y.X. Wang, K. Lenahan, R.O. Claus and J.R. Heflin, *Proc. Am. Chem. Soc.* (Boston), August 1998.
14. W. Ren, et al., *Mater. Lett.* **31**, 185 (1997).
15. Q.M. Zhang, W.Y. Pan and L.E. Cross, *J. Appl. Phys.* **63**, 2492 (1998).
16. Z.J. Zheng, I.L. Guy and T.L. Tansley, *J. Intelligent Mater. System and Struct.* **9**, 69 (1998).

Part II

Photorefractive Polymers

DESIGN AND OPTIMIZATION OF CHROMOPHORES
FOR LIQUID CRYSTAL AND PHOTOREFRACTIVE APPLICATIONS

R. J. TWIEG, M. HE, L. SUKHOMLINOVA, F. YOU
Department of Chemistry, Kent State University, Kent, OH 44242   rtwieg@lci.kent.edu

W.E. MOERNER, M.A. DIAZ-GARCIA, D. WRIGHT, J.D. CASPERSON
Department of Chemistry, UC San Diego, La Jolla, CA 92093-0340

R. WORTMANN, C. GLANIA, P. KRÄMER, K. LUKASZUK, R. MATSCHINER
Institute of Physical Chemistry, University of Mainz, Mainz, Germany

K. D. SINGER, V. OSTOVERKHOV, R. PETSCHEK
Department of Physics, Case Western Reserve University, Cleveland, OH 44106-7079

ABSTRACT

Organic chromophores have been exploited for a wide range of discrete optical and electronic functions as well as a growing number of combined opto-electronic functions. We are pursuing development of organic and polymer materials for a range of applications that require properties including liquid crystallinity, second order optical nonlinearity, photorefractivity and, more recently, special nonlinear optical behavior involving molecular chirality.

INTRODUCTION

Already well known and established applications for organic chromophores include photosensitizers and charge transport agents for electrophotography and liquid crystals and dichroic dyes for displays. New and emerging applications of organic chromophores include those for electro-optic, electroluminescent and photorefractive devices. Fortunately, in most cases, many common structure-function features can be shared amongst these classes of chromophores and as a result many synthetic, design and structure-function relationships are portable amongst them. On the other hand, some newer and presently less well understood chromophore demands arise, such as in the present case of the chiral NLO chromophores.

Our own current efforts on organic chromophores reflect this diversity of structure and function but again attempt to utilize common structure concepts as often as possible. The most predominant overlap in chromophore design we currently pursue is between liquid crystals and both nonlinear optical (NLO) chromophores and photorefractive (PR) chromophores. For example, for both photorefractive and liquid crystal applications high birefringence is required and we are working with chromophores that contain the diphenylacetylene and methylenedihydropyridine functionalities. We are also attempting to extend these existing methylenedihydropyridine systems to the analogous reduced methylenetetrahydro- and methylenehexahydro-systems and to still more condensed polycyclic systems containing these heterocycles with the anticipation of additional beneficial electronic and optical functions.

One ultimate goal here is to prepare and optimize liquid crystal compositions with useful photorefractive function. This evolution comes from a number of established and highly successful polymer PR composite systems that have already been examined. In these polymer composites a wide range of useful chromophores are known along with a complementary set of transport agents, sensitizers and plasticizers. Existing photorefractive materials are generally fabricated and their diffraction efficiency and speed are evaluated with a charge transport material such as PVK. The trick here will be to endow liquid crystals with all of these discrete

Mat. Res. Soc. Symp. Proc. Vol. 561 © 1999 Materials Research Society

functions. Some initial promising work indicates that this is a distinct possibility. Especially interesting is the creation of liquid crystal materials with discrete charge (electron or hole) transport capabilities. Such liquid crystals will be valuable not only for the photorefractive effect but a variety of other electroluminescent devices and the like. The liquid crystal behavior of the chromophores is evaluated by both thermal analysis and by polarized optical microscopy. Many of the relevant electronic and optical properties of chromophores are also conveniently obtained by electro-optical absorption measurements (EOAM). This latter single technique provides simultanously a wealth of information about a range of important chromophore properties.

Chirality is another general feature of current interest as it applies to NLO materials with helical π-systems for Kleinman-disallowed hyperpolarizability. Here the pseudotensor contributions to the nonlinearity are evaluated by hyper-Rayleigh light scattering experiments. Presently the important design features that will give rise to large nonlinearities via these contributions are not well understood. The chiral molecules we are presently evaluating almost invariably are derived from natural sources such as camphor or the steroid estrone. Chirality, of course, has already been of great relevance to liquid crystals including cholesteric and ferroelectric liquid crystals. In all cases we are also attempting to design these molecules mindful of ultimate requirements for durability as well.

PHOTOREFRACTIVE CHROMOPHORES

Since their first demonstration just about a decade ago photorefractive polymers have improved very significantly in overall performance.[1] The main criteria by which photorefractive polymers are evaluated include the diffraction efficiency and the speed of evolution of the photorefractive grating. Of course, there are numerous other material and device parameters that can be critical to the performance of a practical photorefractive device. Some of these issues are general and pervasive while others might be very specific or even unique to a particular device configuration or function. The durability of the photorefractive media and the potential (electric field) required for device operation are examples of more general issues while the grating lifetime requirements might be more specific to one application vs. another (as in dynamic switching vs. optical storage).

Numerous types of photorefractive chromophores and compositions containing them have already been described. A selected menu of structures is found in Figure 1. Many of these polar and conjugated molecules have roots in the antecedent search for second order nonlinear optical chromophores but were later adopted for use as photorefractive chromophores. In fact, many of these chromophores turn out to be better suited for PR applications than they were for NLO applications. This all came as a result of the combined refined understanding of both the basic polarizability properties of NLO chromophores [2] and the basic mechanistic origins of grating formation in low T_g composite PR systems.[3]

The β-nitrostyrene FDEANST 1 [4] is just a fluorinated version of DEANST [5] with less absorption that permitted the first demonstration of net gain. The azodye DMNPAA 2 was the first chromophore utilized in a system with approaching 100% diffraction efficiency.[6] The dicyanostyrenes PDCST 3 and AODCST 4 represent an interesting sequence of chromophore development. The PDCST compound comes from earlier studies as an NLO chromophore but served very well in composite PR structures.[7] The compound AODCST is a highly tuned derivative with thus far the best performance in this class.[8] Other diactivated styrenes also produce interesting materials. The cyanoester CEST-2 5 [9] and the ATOP-1 6 [10] represent efforts to further tune physical properties in the diactivated styrene series. In the latter case of ATOP-1 careful adjustments in both ring systems and acceptors carefully tuned the system to the cyanine limit. Also noteworthy in these compounds are the peripheral aliphatic groups. Although their electronic and optical influence of these aliphatic groups are more subtle (being nonconjugated and optically isotropic) they do play a very important role to adjust the miscibility of the chromophore in the rest of the composition.

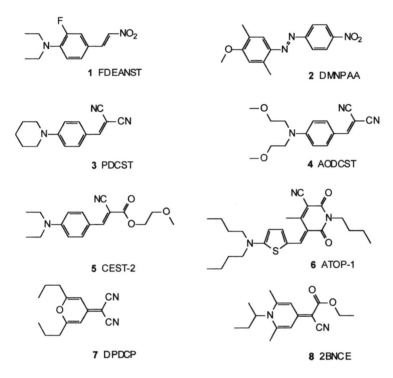

Figure 1. Some Selected Photorefractive Chromophores. These eight structures represent just a few of the numerous chromophores already appearing in the literature that have been utilized for the preparation of organic and polymeric photorefractive compositions.

One simple and distinctive structure class of photorefractive chromophores possess a methylenedihydropyridine structure as in the case of 2BNCE **8**.[11] These chromophores (which themselves evolved from optimization of methylenedihydropyran systems such as DPDCP **7** [12]) were found to possess very large polarizability anisotropies and very large ground state dipole moments which makes them well suited for photorefractive operation involving the so-called "birefringence enhancement" mechanism [3] rather than the conventional electrooptic effect. These chromophores also have interesting glass forming properties that permit the preparation of high optical quality samples. An unsolved drawback of these chromophores when they are used without additional transport agents is their slow grating growth rates.

Our present efforts involve extension of these methylenedihydropyridine chromophores to further enhance their photorefractive function. The mode of enhancement here is simply to prepare and evaluate the vinylogous analogues. Representative compounds **9-14** in this class are found in Figure 2. Here we have the dihydropyridine ring coupled to the acceptors by two double bonds rather than a single double bond. We know these compounds are close to the cyanine limit

121

(here β is near a minimum and α near a maximum, since α is near maximum then anisotropy in that parameter is most feasible). The results of a simple solvatochromic evaluation for compound **11** shows a shift of <30 nm (hexane through nitromethane). Two of these new compounds, along with a number of others found in Figure 1 have been studied by EOAM.[13] Some EOAM data are summarized in Table 1 and a typical spectra is found in Fig. 3. There are a variety of merit figures which have been proposed for photorefractive chromophores.[14,15,16] In this case the merit figure F_0^{Kerr} is utilized (Equations 1-3).[11]

The important salient features here in assessing merit are the distinction between the contributions due to the first hyperpolarizability β_o and the anisotropy in the linear polarizability $\delta\alpha_o$. Molecules cannot be simultaneously fully optimized for both α and β so here we have chosen to just optimize for α (or $\delta\alpha_0$). Another important consideration is that the merit figure scales as the reciprocal of the molecular weight M. It is therefor desirable to work with the smallest possible molecule and this is in and of itself a difficult tradeoff. Generally, a long and conjugated system is required to obtain the electronic properties and in order to make the material tractable one needs to add other functionality. As can be seen from the table, compound **10** has by far the highest figure of merit. This is the result of not only its favorable electronic properties (large ground state dipole μ_g and index anisotropy $\delta\alpha_0$) but also the fact that it has the smallest molecular weight M. The comparable cyanoester **12** has a significantly reduced, albeit still useful, merit figure. This is due to a substantial decrease in ground state dipole moment and increase in molecular weight. The ester acceptor is larger and less effective than the cyano group and also has a degree of freedom of rotation to diminish the ground state dipole.

$$F_0^{Kerr} = [9\mu_g\beta_0 + 2\mu_g^2\delta\alpha_0 /(kT)]/ M \qquad (1)$$

$$\delta\alpha_0 = 2\mu_{ag}^2\lambda_{ag} / (hc_0) \qquad (2)$$

$$\beta_0 = 6\mu_{ag}^2 \Delta\mu \, \lambda_{ag}^2 / (hc_0)^2 \qquad (3)$$

	Chromophore	1	2	3	6	7	8	10	12
	Abbreviation	FDEANST	DMNPAA	PDCST	ATOP	DPDCP	2BNCE		
M	g mol^{-1}	238	285	237	428	228	274	197	244
λ_{ag}	Nm	417	391	422	536	352	378	493	497
μ_g	10^{-30} C m	21	21	29	47	27	31	50	37
$\Delta\mu$	10^{-30} C m	55	44	33	6	1	−9	−5	−2
$\delta\alpha_0$	10^{-40} C V^{-1} m^2	22	22	25	55	16	19	50	46
β_0	10^{-50} C V^{-2} m^3	78	56	52	27	1	−9	−17	−6
F_0	10^{-74} C^2 V^{-2} m^{4}kg^{-1}mol	0.27	0.20	0.48	1.40	0.24	0.30	3.09	1.22

Table 1. Some Physical and Electrooptical Properties. The estimated Kerr figure of merit F_0 of the chromophores and other properties were obtained in dioxane solution.[13]

Unfortunately, compound **10** is difficult to load to a significant extent in the PVK composite without phase separation or crystallization. In order to make it more miscible we have evaluated the influence of additional substituents on this basic structure. The miscibility and processability scales with the melting point (to a first approximation only). As can be seen in the

Fig. 2 modification of the substituents can result in a significant reduction in the melting temperature. The most productive means found thus far to lower the melting point is the well established practice of chain branching. Note that all compounds with the exception of **12** (a cyanoester) are dicyano substituted. The cyanoester compounds usually have lower melting points (but, unfortunately, inferior PR performance) than their dicyano counterparts. Compounds such as those in Fig. 2 are now under active evaluation as photorefractive chromophores.[17] For example, compound **13** has been successfully utilized in a near-IR sensitive composition (Fig. 4).

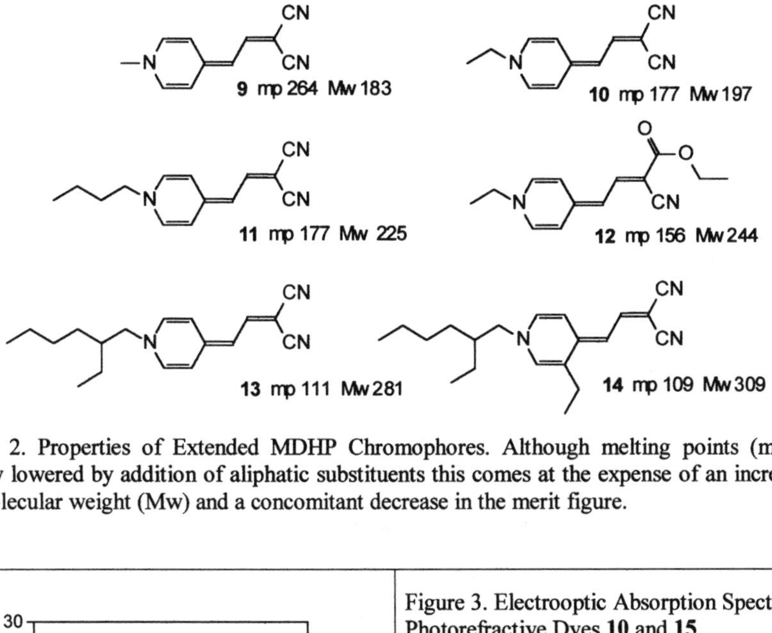

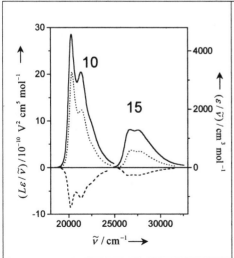

Figure 2. Properties of Extended MDHP Chromophores. Although melting points (mp) are usually lowered by addition of aliphatic substituents this comes at the expense of an increase of the molecular weight (Mw) and a concomitant decrease in the merit figure.

Figure 3. Electrooptic Absorption Spectra of Photorefractive Dyes **10** and **15**

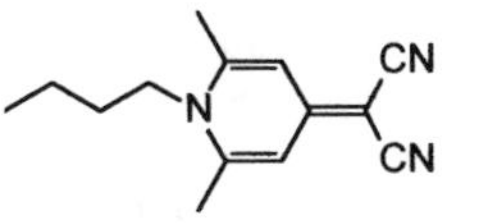

The EOAM spectra for both compounds are governed by electrodichroism: they are nearly mirror images, the band shifts are very small and they also both closely resemble the normal absorption spectra with no external field applied

key: right scale (———) normal absorption
left scale (·············) $\phi = 0°$ E field
left scale (-------) $\phi = 90°$ E field

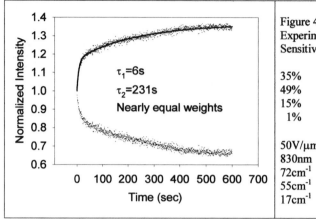

Figure 4. Two Beam Coupling Experiment on a Near IR Sensitive PR Composition

35%	Chromophore **13**
49%	Polyvinylcarbazole
15%	Butylbenzylphthalate
1%	TNFDM

50V/μm	applied field
830nm	wavelength
72cm⁻¹	two beam coupling gain
55cm⁻¹	adsorption coefficient
17cm⁻¹	net gain

(Graph: Normalized Intensity vs Time (sec), with τ_1=6s, τ_2=231s, Nearly equal weights)

Merging Liquid Crystal and Photorefractive Chromophore Structure

Currently it is of great interest to identify liquid crystalline photorefractive materials. Two levels of activity are underway. The first level exploits known liquid crystals for their birefringence and plasticizing properties as chromophores in composite media.[18] The second level examines photorefractivity in the liquid crystalline state [19] and even the polymer modified (PDLC) liquid crystalline state.[20]

We have just discussed some photorefractive chromophores based on methylene dihydropyridines and their vinylogous counterparts. As an extension of this effort we hope to introduce these kinds of structures into liquid crystals. For example, in Fig. 5, we see the extraction of structure features from a pair of chromophores to give a general hybrid structure with possible simultaneous liquid crystalline and photorefractive / NLO utility. Liquid crystals such as **16** are already well known [21,22] and the pentahydronapthalene **17** is demonstrated to be a very good photorefractive chromophore.[23] What are the potential benefits of this hybridization? First, aryl amine donors are presently preferred if stability is an issue.[24] Second, if olefins are part of a conjugated system of a dye they are best kept endocyclic for stability enhancement. Third, inclusion of the donor atom in the ring assists maximal overlap and also potential (but as yet unproven) thermochemical or photochemical stabilization as well.

We have not yet prepared the hybrid chromophore found in Fig. 5 but all indications are that it should be possible to make. In preliminary studies already underway we are first examining mononuclear systems in the respective series of dihydropyridine, tetrahydropyridine and hexahydropyridine. The series is not yet complete but fragments of data are now available as mesomorphism is demonstrated so far only in a group of materials without a common core group. Compounds in this new series are found in Figure 6. The dihydropyridine compound **18** is representative of the many liquid crystals now known with this structure. We have subsequently found liquid crystallinity in the analogous set of tetrahydropyridines such as **19**. The hexahydropyridine set is just now being examined and compound **20** is not liquid crystalline. However, as a proof of principle, mesogenic activity has been identified in a more extended benzoate ester-containing molecule **21**. Efforts on a contiguous series with a larger common core group are underway which will permit more meaningful comparisons of liquid crystallinity and other electronic and optical properties.

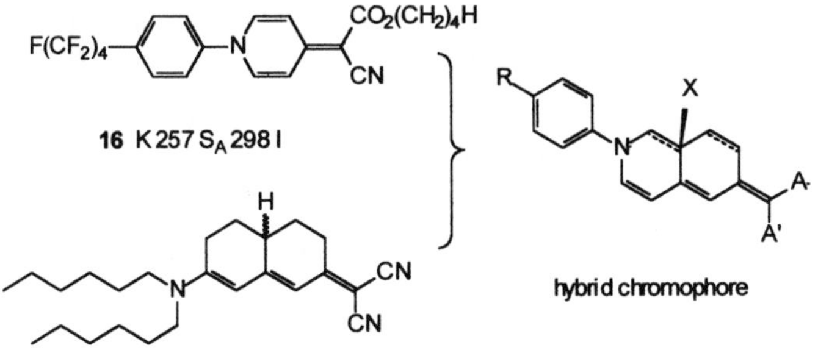

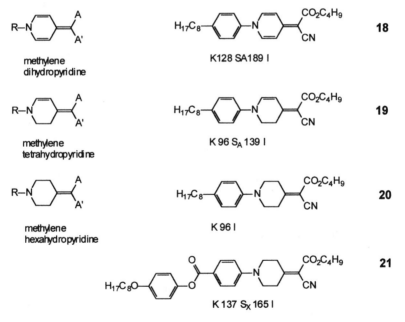

16 K 257 S$_A$ 298 I

17 DHADC-MPN

hybrid chromophore

Figure 5. Chromophore "Hybridization". Compound **16** is a known liquid crystal [22] and compound **17** is a known photorefractive chromophore.[23] By selecting appropriate substructures from each compound it should be possible to prepare new chromophores with some combined properties. Substituent R can be other additional conjugation or donating groups, if X is present it can be a stereogenic center (but note that the --- to the bridgehead conjugation is not possible), if X is nil then the conjugated segments (--- — ---) can be either present or absent.

methylene
dihydropyridine

18

K128 SA189 I

methylene
tetrahydropyridine

19

K 96 S$_A$ 139 I

methylene
hexahydropyridine

20

K 96 I

21

K 137 S$_X$ 165 I

Figure 6. Liquid Crystals with Heterocycle Cores. As the heterocycle is gradually chemically reduced more nonplanar conformations become possible and mesogenic activity diminishes. This iterative change in conjugation pathway also influences the overall electronic properties as well.

Some liquid crystals have already been prepared which look very much like well-known NLO chromophores. As a general strategy we are attempting to build liquid crystals that have a maximum of π-content and a minimum of aliphatic content (or even *no* aliphatic component whatsoever). The motivations for minimizing aliphatic content are the same as were discussed before for the PR chromophore design, i.e., to maximize the content of the active conjugated component. In order to accomplish this goal we have followed up on some earlier findings involving the influence of semifluorination [25] on mesogenic activity in organic compounds. Simple hydrocarbons (n-alkanes) have no mesomorphic behaviour at all but if one selectively incorporates a fluorinated block into a hydrocarbon (compare **22** and **23** in Fig. 7) mesogenic behaviour can result.[26]

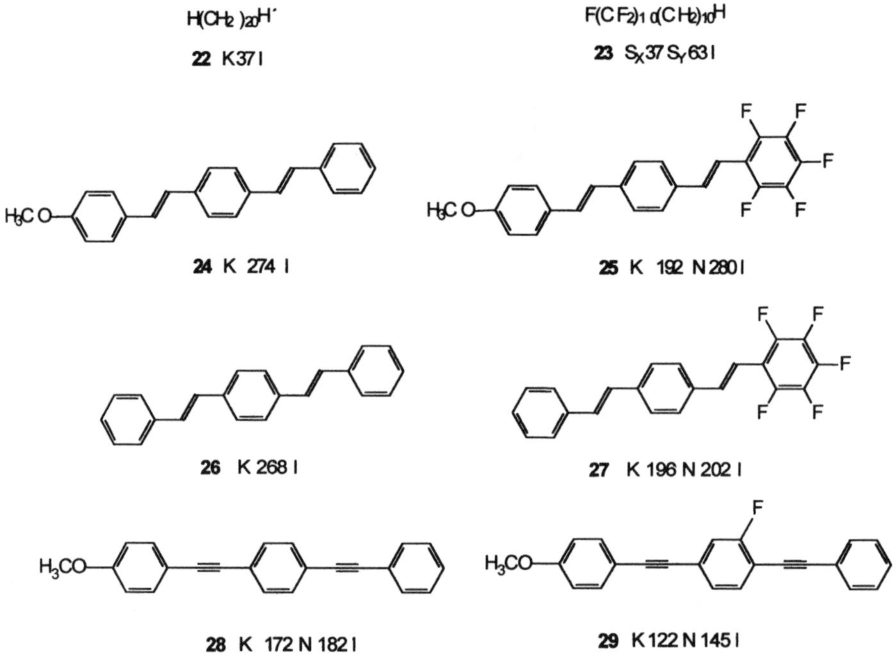

H(CH₂)₂₀H·

22 K 37 I

F(CF₂)₁₀(CH₂)₁₀H

23 S$_X$ 37 S$_Y$ 63 I

H₃CO

24 K 274 I

H₃CO

25 K 192 N 280 I

26 K 268 I

27 K 196 N 202 I

H₃CO

28 K 172 N 182 I

H₃CO

29 K 122 N 145 I

Figure 7. The Influence of Fluorination on Mesogenicity. The C20 hydrocarbon eicosane **22** is not mesogenic but the analogue **23** with a C10 fluorinated block has smectic (G and/or J) behaviour. The methoxy bis stilbene **24** is not a mesogen but the analogue **25** with a pentafluorinated ring has a large nematic range. Likewise, parent stilbene **26** is not mesogenic but the pentafluorinated analogue **27** is a nematic liquid crystal. The monofluorinated methoxy bistolane **29** has lower mesophase temperatures than nonfluorinated analog **28**. It is clear that the introduction of fluorinated units is a useful tool to influence mesogenic activity in both aliphatic and aromatic liquid crystals.

The question now arises ... can we accomplish enhancement of mesogenic activity by selective fluorination in aromatic systems as well? In fact, there is already a hint that this could be the case. In a study of the incorporation of fluorine into stilbenes and the influence of fluorination on their hyperpolarizability mesogenic activity was discovered in some highly locally fluorinated derivatives (compare **24** and **25**).[27] We have also found that fluorinated parent bis stilbene is mesogenic while nonfluorinated analog [28] is not (compare **26** and **27**). In this fashion mesogenic activity has been introduced without significant dilution of the interesting electronic and optical properties by introduction of any aliphatic groups. Such efforts are far from over, however, because it will be important to move the high mesophase temperatures of these compounds closer to ambient and to increase the temperature window of mesogenic phase. This may require selective fluorination elsewhere in the molecule or judicious use of a minimal amount of aliphatic component. For example, we have subsequently introduced fluorinated structure features into bistolanes that contain a single tail. The incorporation of one fluorine atom into the middle ring of methoxy bistolane has been shown to decrease melting and clearing temperatures significantly (compare **28** and **29**). The effects of fluoro substitution at different positions in a variety of core systems have been already discussed. [29]

Chiral Chromophores

Most second-order nonlinear optics in organic materials has been aimed at understanding and exploiting dipolar molecules containing delocalized electrons with electron withdrawing and/or donating substituents.[30] These compounds possess large molecular hyperpolarizabilities that have been utilized in polar macroscopic materials such as molecular crystals, poled polymers, and monomolecular films.[31] These compounds often have a one-dimensional character, and, as such, necessarily satisfy Kleinman (full-permutation) symmetry.

The molecular hyperpolarizability, however, has other components of interest. The hyperpolarizability, a third-rank tensor, describes the relevant nonlinear optical properties. This tensor can be decomposed into irreducible tensors and associated scalar figures of merit. A rank-one tensor, which transforms as a vector, describes the polar response described above. Its figure of merit can be measured using EFISH. In the Kleinman symmetric regime, this vector and a third rank tensor, the octupolar component, fully describe the response. This octupolar component has been the subject of recent research, and molecular properties for efficient octupolar interactions have been elucidated.[32,33,34]

We have recently discussed an experiment which measures all scalar invariants of the SHG molecular hyperpolarizability in the Kleinman-disallowed regime using hyper-Rayleigh scattering (HRS).[35] Including Kleinman symmetry breaking, there are four irreducible tensors corresponding to six scalar figures of merit represent the SHG hyperpolarizability.[36] One of these is a second rank pseudotensor called the quadrupolar component. In a chiral, uniaxial system (such as a liquid crystal or stretched polymer) such a hyperpolarizability will result in a non-zero non-linear optical susceptibility. Like the octupolar susceptibility this component necessarily involves electrons which move in at least two dimensions. It is possible that the presence of strong donors and acceptors will increase this hyperpolarizability. Other than these facts / speculations little is known about the structural features of a chromophore which will result in large Kleinman disallowed quadrupolar hyperpolarizabilities.

Measurement of all the Kleinmann allowed hyperpolarizabilities requires examining the amplitude of HRS for several combinations of incoming and outgoing polarizations. We have improved our measurement technique using a better choice of polarizations. We then measured or re-measured several known compounds and new chiral compounds. All these compounds are expected to have electronic motion in more two or three dimensions. They have differing strengths of donors and acceptors. Some have electrons delocalized in two dimensions. Some have electrons delocalized in a chiral manner in three dimensions.

Material	$\beta^2_{1mm}/\beta^2_{1ss}$	$\beta^2_{1sm+}/\beta^2_{1ss}$	$\beta^2_{1sm-}/\beta^2_{1ss}$	$\beta^2_{2mm}/\beta^2_{1ss}$	$\beta^2_{3ss}/\beta^2_{1ss}$	$\beta^2_{1ss}/(\beta^2_{1ss})_{pNA}$	λ, nm
30	0.12±0.03	-0.02±0.02	0.17±0.1	0.04±0.02	0.52±0.02	1	1064
31	0.27±0.04	-0.15±0.02	0.09±0.02	0.14±0.04	0.51±0.02	~3.3	1064
32	1.4±0.1	-1.44±0.08	0.9±0.5	0.7±0.1	1.56±0.08	~28	1064
33	0.24±0.02	0.09±0.1	-0.05±0.01	0.06±0.02	0.52±0.01	~8	1064
34	0.04±0.03	0.03±0.01	0.03±0.02	0.17±0.03	0.47±0.01	~42	1340
33	0.16±0.04	-0.03±0.02	0.15±0.03	0.00±0.04	0.49±0.02	~3	1340
35	0.12±0.02	0.12±0.01	0.03±0.01	0.00±0.02	0.39±0.01	~114	1340

Table 2. Results of hyper-Rayleigh scattering experiments. All entries are scaled to the magnitude of the vector (1ss) component.

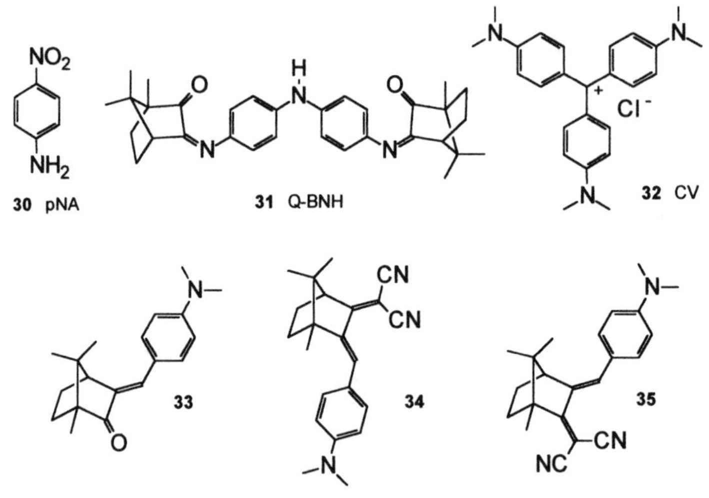

Figure 8. Compounds Studied by hyper-Rayleigh Scattering. In the chiral molecules with stereogenic centers (31, 33, 34, 35) the extent of handed helical twist in the π-system may influence the magnitude of the 2mm component

Table 2 contains the results of measurements of the scalar invariants of several chromophores shown in Figure 8. The 1mm, 1sm+, and 1sm- are all vector components of various symmetry, while the 2mm and 3ss are the quadrupolar and octupolar components.[35] The 1ss component is scaled to the vector component of p-NA (30) which has been measured to be about 12 x 10^{-30} esu.[37] As expected, the Kleinman symmetry breaking components of p-NA are negligible. The larger accepting strength in **34** results in a substantially larger 2mm (quadrupolar) component both relative to 1ss and in absolute terms compared with **33**, which does not exhibit significant 2mm. X-Ray measurements also indicate that the twist angle of conjugated π-system in molecule **34** is largest (11°) compared with **33** (1.8°) or **35** (1.4°).

The acceptor-donor-acceptor structure of Q-BNH (**31**) results in a significant 2mm component (all compared to the 1ss component). Crystal violet (CV, **32**) exhibits very large 2mm and 3ss components, but does have a significant 1ss component, which must arise from a polar conformation in solution. The observation of 1ss and 2mm components in crystal violet indicates that the D_3 symmetry is broken in the ground state in solution [32]. The donor-acceptor-donor nature also seems to contribute to a large 2mm component.

One can conclude that multidimensional nature and strong donors and acceptors lead to large 2mm components. Further work is necessary to gain of better understanding of the qualities needed for enhancing the quadrupolar figure of merit, as well as to exploit this nonlinear susceptibility in a bulk material.

Acknowledgements

Support for this work at Kent State University and Case Western Reserve University from ALCOM (NSF DMR 89-20147) is gratefully acknowledged.

Support of this work at UC San Diego (present address: Chemistry Department, Stanford University) from AFOSR (F49620-96-1 and F49620-97-1-0286) is gratefully acknowledged.

Support for this work at Mainz University (present address: University of Kaiserslautern) from the Stiftung Volkswagenwerk and the Fonds der Chemischen Industrie is gratefully acknowledged.

We thank Dr. Charles Barnes at University of Missouri-Columbia for x-ray structure determination of compounds **34** and **35**.

References

[1] W. E. Moerner, A. Grunnet-Jepsen, C. L. Thompson, *Ann. Rev. Mater. Sci.*, **27**, 585 (1997).

[2] S. R. Marder, C. B. Gorman, B. G. Tieman, L.-T. Cheng, *J. Amer. Chem. Soc.*, **115**, 3006 (1993).

[3] W. E. Moerner, S. M. Silence, F. Hache, G. C. Bjorklund, *J. Opt. Soc. Am. B*, **11**, 320 (1994).

[4] M. C. J. M. Donckers, S. M. Silence, C. A. Walsh, F. Hache, D. M. Burland, W. E. Moerner, R. J. Twieg, *Opt. Lett.*, **18**, 1044 (1993).

[5] M. E. Orczyk, J. Zieba, P. N. Prasad, *J. Phys. Chem.*, **98**, 8699 (1994).

[6] K. Meerholz, B. L. Volodin, Sandalphon, B. Kippelen, N. Peyghambarian, *Nature*, **371**, 497 (1994).

[7] A. Grunnet-Jepsen, C. L. Thompson, R. J. Twieg, W. E. Moerner, Appl. *Phys. Lett.*, **70**, 1515 (1997).

[8] D. Wright, M. A. Diaz-Garcia, J. D. Casperson, M. DeClue, W. E. Moerner, R. J. Twieg, *Appl. Phys. Lett*, **73**, 1490 (1998).

[9] M. A. Diaz-Garcia, D. Wright, J. D. Casperson, B. Smith, E. Glazer, W. E. Moerner, L. I. Sukhomlinova, R. J. Twieg, *Chem. Mater.*, submitted.

[10] F. Wurthner, R. Wortmann, R. Matschiner, K. Lukaszuk, K. Meerholz, Y. DeNardin, R. Bittner, C. Brauchle, R. Sens, *Ang. Chem. Int. Ed.*, **36**, 2756 (1998).

[11] P. M. Lundquist, R. Wortmann, C. Geletneky, R. J. Twieg, M. Jurich, V. Y. Lee, C. R. Moylan, D. M. Burland, *Science*, **274**, 1182 (1996).

[12] R. Wortmann, C. Poga, R. J. Twieg, C. Geletneky, C. R.. Moylan, P. M. Lundquist, R. G. DeVoe, P. M. Cotts, H. Horn, J. E. Rice, D. M. Burland, *J. Chem. Phys.*, **105**, 10637 (1996).

[13] R. Wortmann, C. Glania, P. Krämer, K. Lukaszuk, R. Matschiner, R. J. Twieg, F. You, *Chem. Phys.*, accepted for publication.

[14] R. Wortmann, C. Poga, R.J. Twieg, C. Geletneky, C.R. Moylan, P.M. Lundquist, R.G. DeVoe, P.M. Cotts, H. Horn, J. Rice, D.M. Burland, *J. Chem. Phys.* **105** (1996) 10637.

[15] B. Kippelen, F. Meyers, N. Peyghambarian, S. R. Marder, *J. Amer. Chem. Soc.*, **119**, 4559 (1997).

[16] C.R. Moylan, R. Wortmann, R.J. Twieg, I.-H. McComb, *J. Opt. Soc. Am. B* **15** (1998) 929.

[17] D. Wright, W. E. Moerner, F. You, R. J. Twieg, manuscript in preparation.

[18] J. Zhang, K. D. Singer, *Appl. Phys. Lett.*, **72**, 2948 (1998).

[19] G. P. Wiederrecht, M. R. Wasielewski, *J. Amer. Chem. Soc.*, **120**, 3231 (1998).

[20] A. Golemme, B. L. Volodin, E. Kippelen, N. Peyghambarian, *Opt. Lett.*, **22**, 1226 (1998).

[21] D. J. Dyer, V. Y. Lee, R. J. Twieg, *Liq. Cryst.*, **24**, 271 (1996).

[22] F. You, R. J. Twieg, D. Dyer, V. Lee, *Mol. Cryst. Liq. Cryst*, accepted for publication.

[23] B. Kippelen, S. R. Marder, E. Hendrickx, J. L. Maldonado, G. Guillemet, B. L. Volodin, D. D. Steele, Y. Enami, Sandalphon, Y. J. Yao, J. Wang, H. Rockel, L. Erskine, N. Peyghambarian, *Science*, **279** 54 (1998).

[24] R. B. Prime, G. Y. Chiou, R. J. Twieg, *J. Thermal. Anal.*, **46**, 1133 (1996).

[25] M. Napoli, *J. Fluorine Chem.*, **79**, 59 (1996).

[26] C. Viney, T. Russell, L. Depero, R. Twieg, *Liq. Cryst.*, **5**, 1783 (1989).

[27] C. R. Moylan, K. M. Betterton, R. J. Twieg, C. A. Walsh, *Nonlinear Opt.*, **8**, 69 (1994).

[28] T.W. Campbell, R.N. McDonald, *J. Org. Chem.* **24**, 1246 (1959).

[29] M. Hird, K. Toyne, *Mol. Cryst. Liq. Cryst.* **323**, 1 (1998).

[30] J. Zyss, ed., *Molecular Nonlinear Optics* (Academic Press, 1994).

[31] G. Lindsay and K. Singer, eds., *Polymers for Second-Order Nonlinear Optics* (ACS Symp. Ser. 601, 1995).

[32] J. Zyss and I. Ledoux, *Chem. Rev.* **94**, 77 (1994).

[33] S. Brasselet and J. Zyss, *J. Opt. Soc. Am. B.* **15**, 257 (1998).

[34] M. Joffre, D. Yaron, R. Silbey, and J. Zyss, *J. Chem. Phys.* **97**, 5607 (1992).

[35] S.F. Hubbard, R.G. Petschek, K.D. Singer, N. D'Sidocky, C. Hudson, L.C. Chien, C.C. Henderson, and P.A. Cahill, *J. Opt. Soc. Am. B* **15**, 289 (1998).

[36] M. Kozierowski, *Phys. Rev. A* **31**, 509-510 (1985); G. Wagniere, *Appl. Phys. B* **41**, 169 (1986); R. Wortmann, P. Krämer, C. Glania, S. Lebus, and N. Detzer, *Chem. Phys.* **173**, 99 (1992).

[37] K.D. Singer, J.E. Sohn, L.A. King, H.M. Gordon, H.E. Katz, and C.W. Dirk, *J. Opt. Soc. Am. B.* **6**, 1339 (1989).

PHOTOREFRACTIVE POLYMERS WITH HIGH SPEED

N. Peyghambarian, K. B. Ferrio[†], J. A. Herlocker, E. Hendrickx, B. D. Guenther, S. Mery[*], Y. Zhang[**] and B. Kippelen[†]

Optical Sciences Center, University of Arizona, Tucson
[*] Institut de Physique et Chimie des Matériaux de Strasbourg, Groupe des Matériaux Organiques, CNRS, Strasbourg, France
[**] Department of Chemistry, University of Arizona, Tucson

ABSTRACT

Rapid progress in photorefractive polymers includes new materials with improved response rates. We briefly review recent developments in this rapidly evolving area and report grating formation with a 2-millisecond time-constant, the fastest reported to date in any photorefractive polymer.

INTRODUCTION

The discovery [1] of photorefractivity, a nonlocal nonlinear optical effect by which energy may be coherently transferred between two or more optical beams, and the subsequent recognition of potential applications [2] continue to attract strong interest more than three decades hence. The bulk of this research remains concerned principally with inorganic crystalline materials, but promising new vistas were opened in 1990, by the first observation of photorefractivity in a polymer composite [3]. The potential advantage of polymers over photorefractive crystals is that the essential requirements of photorefractivity can be controlled by doping a conducting polymer host: efficient photogeneration can be tailored with an appropriate sensitizer; traps can be incorporated in concentrations large compared to acceptor densities in crystals; and the electro-optic response is provided by molecules engineered to purpose. The promise of inexpensive materials with tailored compositions and amenable to commercial production immediately spawned intense interest in applications and basic research alike.

Although the efficiency of early materials was severely limited by low molecular hyperpolarizabilities, the incorporation of optically anisotropic moieties with permanent dipole moments orientable in the dynamic space-charge field enabled thin holograms with useful diffraction efficiencies [4] and soon led to composites with near 100% diffraction efficiency [5]. Effective orientation of the birefringent species requires composites with low glass-transition temperatures, and much effort was invested in developing low-T_g materials with high net efficiency and good stability. The response-times of these materials typically ranged from several tens of milliseconds to minutes; only recently have two-beam coupling response-times of 7.5 ms [6] and 5 ms [7] been reported. Our group has recently reported hologram-formation times of 4 ms [8]. Here, we discuss our latest observation of a 2-ms hologram-formation and emphasize the important distinction between grating formation and two-beam coupling or four-wave mixing response times.

[†] corresponding authors: kyle@u.arizona.edu, kippelen@u.arizona.edu

THEORY

The two principal optical measurements of interest are two-beam coupling, a key signature of photorefractivity, and diffraction efficiency. Two coherent 633-nm HeNe laser beams of intensities I_1 and I_2 are made to overlap and interfere in the polymer film, as shown in Figure 1.

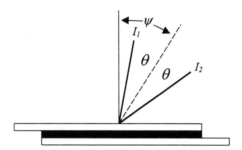

Figure 1: Geometry for writing an holographic photorefractive grating. The angles are measured in air; refraction at the interfaces is not shown, for simplicity.

An electric bias of several tens of volts per micron is applied across the sample, via the ITO-coated slides. This field is necessary, because both the photogeneration efficiency and carrier mobility in polymers are strongly field-dependent. Carriers are then generated in a pattern with the spatial periodicity of the intensity grating. Although electrons are effectively bound to the sensitizer moieties, holes that escape geminate recombination diffuse through the hopping manifold provided by PVK and ECZ. The resulting separation of holes from electrons produces a space-charge field, which evolves to an electrostatic equilibrium prescribed by the self-consistent solution of photogeneration, transport, and the Poisson equation. The birefringent polar chromophores reorient partially in the space-charge field, giving rise to a periodic macroscopic orientation. The sample is tilted to obtain a non-vanishing effective index modulation amplitude, Δn. The two-beam coupling gain coefficient is given by

$$\Gamma = (4\pi / \lambda)\Delta n \sin\phi, \tag{1}$$

where ϕ is the phase shift between the optical intensity pattern and the index grating, and λ is the optical wavelength in vacuum. Two-beam coupling is typically described by the experimentally observed gain,

$$\gamma = \left.\frac{I_2(I_1 \neq 0)}{I_2(I_1 = 0)}\right|_{\substack{OUTPUT \\ PLANE}} = \frac{(1+m)\exp(\Gamma L)}{m + \exp(\Gamma L)}, \quad m = \left.\frac{I_1}{I_2}\right|_{\substack{INPUT \\ PLANE}}. \tag{2}$$

Diffraction efficiency, by contrast, depends only on the index modulation. For this experimental configuration, the internal diffraction efficiency (corrected for weak absorption) of grating spacings up to about 5 μm are reasonably described by Kogelnik's well known expression for thick gratings [9]:

$$\eta_{INT} \approx \sin^2\left(\frac{\pi\Delta nd}{\lambda\sqrt{c_R c_S}}\right),$$
(3)

with sample thickness d and the familiar direction cosines c_R and c_S defined by Kogelnik [9]. Since diffraction efficiency depends only on the index modulation, and not on the phase-shift of the grating, this work has focused on using diffraction measurements to understand the temporal evolution of the index modulation as an essential first step in describing the full photorefractive dynamics. For comparison, we note that Wright *et al.* have recently reported a dominant 4- to 5-ms component [7] for γ. The composite we describe here has time-constant of 2 ms for Δn under conditions of grating period, electric field, and total intensity very similar to those reported in [7]. We note that, for a given material, the response time measured for γ in two-beam coupling is generally faster that the response of the index-modulation, Δn. The index-modulation therefore provides a conservative characterization of photorefractive dynamics.

RESULTS

The Composite

Polymer composites were prepared using a mixture of poly-N-vinylcarbazole (PVK, 56 wt.%) and N-ethylcarbazole (ECZ, 28 wt.%) as the host matrix. The hole-transport properties of PVK are well known, and the ECZ functions as a compatible plasticizer to lower the glass transition temperature. The chromophore used in this study was a fluorinated cyano-tolane derivative (FTCN, 15.2 wt.%) with negligible optical absorption throughout the visible spectrum, as shown in Figure 2. The absorption peak is centered at 342 nm, leaving a broad optical window. Transparency is a highly desirable feature, as it provides optimum flexibility in the choice of operating wavelengths. The fluorination of this "push-pull" chromophore is designed to enhance the ground-state dipole moment as well as its birefringence. Both are required to ensure a strong macroscopically averaged birefringence in the photorefractive space-charge field. Photosensitization was provided by (2,4,7-trinitro-9-fluorenylidene)malonodinitrile (TNFDM, 0.8 wt.%).

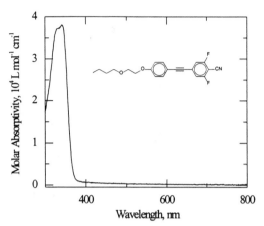

Figure 2: Molar absorptivity of the chromophore (inset) at 10^{-5}M in chloroform.

Samples of the polymer composite were prepared in films of nominally 105-μm thickness, using glass spacers, between borosilicate slides coated with electrodes of indium tin oxide (ITO). Several samples were deliberately delaminated for the purpose of measuring the refractive index of the film; a Metricon 2010 prism coupler operating in substrate mode found $n = 1.68$ at 633 nm. The glass-transition temperature was estimated to be 39°C with modulated differential scanning calorimetry, using a TA Instruments DSC 2920. Samples have exhibited excellent stability for over six months at room temperature and have yet to show any degradation of performance.

Steady-State Measurements

Before making any transient measurements, steady-state measurements were made to confirm the presence of the requisite properties for photorefractivity, with $\psi = 60°$ and $\theta = 10°$. The grating period was approximately 3.1 μm. The diffraction efficiency produced by two s-polarized writing beams, each producing an irradiance of 0.25 W/cm^2 in the sample, and measured with a 5-μW p-polarized probe beam is shown in Figure 3a. The internal diffraction efficiency at 95 V/μm was 24%.

Steady-state two-beam coupling with p-polarized beams, in the same geometry but with $I_1/I_2 = 6.3$ and total irradiance of 0.5 W/cm^2 in the sample, is shown in Figure 3b and reached a value of 34 cm^{-1} at 95 V/μm. The absorption of TNFDM (95 cm^{-1}) results in net loss, but this is not particularly relevant to the determination of index-modulation dynamics.

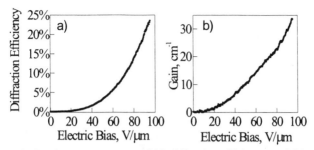

Figure 3: Steady-state responses of (a) diffraction efficiency, and (b) two-beam coupling.

Transient Measurements

The time-evolution of the grating formation was measured in a well controlled sequence to minimize variations of "pre-conditioning" on the samples. First, the electric bias was applied to the sample with only the weak probe beam present. Next, a single writing beam (I_1) was applied. The second writing beam (I_2) was applied 200 ms later, and the formation of the grating was monitored via the evolution of the diffracted intensity. The results in Figure 4 are typical.

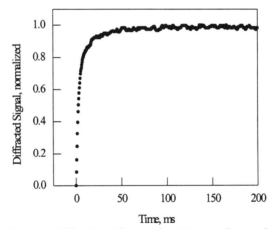

Time, ms

Figure 4: Transient diffraction efficiency for 1.7 μm grating spacing and 95 V/μm bias. An index-modulation response-time of 1.8 ms is inferred from Equations 3 and 4.

In the absence of detailed knowledge of the field-dependent constituent parameters controlling photogeneration and transport, the choice of temporal form for the index modulation (3) is somewhat arbitrary, and fits should be regarded as merely descriptive. Although not obvious by inspection, a simple exponential form for index modulation fails badly. There is some physical motivation for a stretched exponential, but the evidence is not compelling and in practice such fits seldom differ appreciably from a biexponential over the limited time intervals usually studied. We choose the biexponential form for these reasons and to facilitate comparison with the literature:

$$\Delta n = \Delta n_o \left(1 - a \exp(-t / \tau_1) - (1 - a)\exp(-t / \tau_2)\right). \tag{4}$$

This choice gives a dominant, initial time-constant, $\tau_1 = 1.8$ ms, with weight $a = 0.66$. The balance of the response is described by $\tau_2 = 20$ ms. From the perspective of device applications, it may be useful to consider the initial rise time of diffraction efficiency (3) or gain (2) directly. We will not do that here, choosing instead to focus on the underlying fundamental dynamics of Δn. We will only emphasize that the evolution of the gain is much faster yet, as a direct consequence of the exponential dependence in (2), and that the present measurement of a ~2-ms response is faster than any index modulation previously reported. Unfortunately, the low gain of this composite has so far prevented detailed measurements of two-beam coupling transients which remain under investigation.

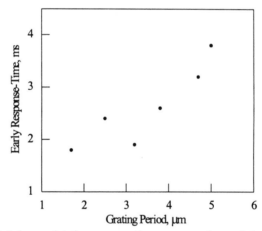

Figure 5: Index-modulation response-time versus grating period.

Additional insight can be obtained by studying the dependence of the index-modulation response-time on grating period, shown in Figure 5. The exact form of this curve is complicated by the field-dependence of the composite properties, but reaches $\tau_1 = 1.8$ ms at a grating spacing of 1.7 μm, a limit imposed by our experimental geometry. This is to the best of our knowledge the fastest index modulation ever observed in a photorefractive polymer composite. The data points in Figure 5 represent repeated measurements with uncertainties of order 0.1 ms; the deviations from a "smooth" line are quite repeatable but not fully understood.

A recent report of a fast photorefractive polymer [7] used the intensity dependence of the response time to show indirectly that, in a specific fast composite, the response-time was limited by photoconduction, not by chromophore orientation. That study did not, however, directly measure the chromophore orientation dynamics and therefore could not project how much faster the composite might be made, given arbitrary hypothetical improvements in photoconductivity. Since the photorefractive response time is fundamentally constrained by the orientational dynamics, knowledge of this limit is an important guide to the appropriation of development efforts. Before turning our attention to improving mobility or photogeneration, we first made direct measurements of the orientational response-time using the simple ellipsometric apparatus of Figure 6.

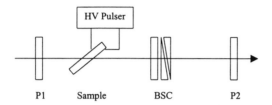

Figure 6: Experimental arrangement for measuring orientational dynamics. P1 and P2 are crossed polarizers at ±45° in the laboratory frame. The Babinet-Soleil compensator BSC nulls stress-induced birefringence in the sample and permits operation in the linear regime of the transmission vs. phase-shift characteristic.

The sample, rotated 45° with respect to the incident laser beam, introduces a phase-shift between the crossed polarizers, P1 and P2. The Babinet-Soleil compensator is adjusted to "bias" the transmission half-way between the peak and the null of transmission. For small birefringence induced by an applied electric field, the transmission is a linear function of the change in index. If the voltage is switched on suddenly, the transmission transient records the time-evolution of the index change. Although this geometry is not exactly that of the diffraction measurements, we believe that by operating at relatively low fields of 6 V/μm, we make a conservative estimate of the orientational response-time.

A typical transient ellipsometric response is shown in Figure 7. If the response is fit to a biexponential function over several hundred milliseconds, a dominant time-constant of approximately 1 ms is obtained. But a hierarchy of time-scales, down to at least tens of microseconds is also observed (not shown). Fitting over a 50-ms interval, during which time the 2-ms response-time diffraction efficiency in Figure 4 achieves 93% of its final value, gives an orientational time-constant of 490 μs. This comparison is underscored in Figure 8, which recapitulates the actual diffraction efficiency of Figure 4, together with that which would be expected if chromophore reorientation in Figure 7 were limiting the response-time.

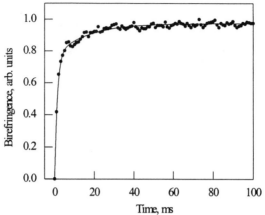

Figure 7 Orientational Birefringence step-response to a 600-V step function.

The orientational limit is significantly faster, by at least 50%, than the fastest actual response measured. The orientational mobility is not limiting the response of the composite, and kilohertz-regime responses may be achievable with improved photoconductivity. Work toward this goal continues.

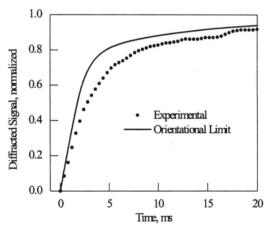

Figure 8: The observed diffraction efficiency lags the orientational limit on all time-scales, up to 100 ms.

CONCLUSIONS

We have demonstrated a very fast photorefractive polymer composite, using a new fluorinated cyano-tolane chromophore in an otherwise conventional photosensitized hole-transporting host to achieve video-rate performance. We have further shown that the unprecedented index-modulation response-time of 1.8 ms is not orientationally limited. Therefore, current work is directed at identifying the influence of the new chromophore on the space-charge field dynamics. The feasibility of further enhancing the dynamic range and gain of these composites in also under active investigation.

ACKNOWLEDGMENTS

This work has been supported by ONR through the MURI Center for Advanced Multifunctional Nonlinear Optical Polymers and Molecular Assemblies (CAMP), AFOSR, NSF, an international collaboration with NSF/CNRS, and a NATO Travel Grant. E.H. acknowledges the support of the Fund for Scientific Research–Flander (Belgium).

REFERENCES

1. A. Ashkin, G.D. Boyd, J.M. Dziedzic, R.G. Smith, A.A. Ballmann, H.J. Levinstein, and K. Nassau, Applied Physics Letters 9, 72 (1966).
2. F.S. Chen, Journal of Applied Physics 38, 3418 (1967).
3. S. Ducharme, J.C. Scott, R.J. Twieg, and W.E. Moerner, Physical Review Letters 66, 1846 (1991).
4. M.C.J.M. Donckers, S.M. Silence, C.A. Walsh, F. Hache, D.M. Burland, and W.E. Moerner, Twieg, R.J., Optics Letters 18, 1044 (1993).
5. K. Meerholz, B.L. Volodin, Sandalphon, B. Kippelen, and N. Peyghambarian, "A photorefractive polymer with high gain and diffraction efficiency near 100%," Nature 371 (6 October), 497-500 (1994).
6. K. Ogino, T. Nomura, T. Shichi, S.-H. Park, H. Sato, T. Aoyama, and T. Wada, Chemistry of Matrials 9, 2768-2775 (1997).
7. D. Wright, M.A. Diaz-Garcia, J.D. Casperson, M. DeClue, W.E. Moerner, and R.J. Twieg, "High-speed photorefractive polymer composites," Applied Physics Letters 73 (11), 1490-1492 (1998).
8. J.A. Herlocker, K.B. Ferrio, E. Hendrickx, B.D. Guenther, S. Mery, B. Kippelen, and N. Peyghambarian, "Direct observation of orientation limit in a fast photorefractive polymer composite," Applied Physics Letters (1999, accepted).
9. H. Kogelnik, "Coupled Wave Theory for Thick Hologram Gratings," The Bell System Technical Journal 48 (November), 2909-2947 (1969).

Part III

Electronic and
Light-Emitting Materials

TIME-RESOLVED TRANSPORT OF ELECTRONS AND HOLES IN CONJUGATED POLYMERS

D. PINNER, R. H. FRIEND, N. TESSLER
Cavendish Laboratory, University of Cambridge, Madingley Road, Cambridge, CB3 0HE, UK

ABSTRACT

Detailed experimental and theoretical analysis of the pulsed excitation of polymer light emitting diodes is presented. We find a set of universal transient features for different device configurations which can be reproduced using our phenomenological numerical model. We find that the temporal evolution of the electroluminescence can be characterised by five main features: i) a delay followed by; ii) fast initial rise at turn-on followed by; iii) a slow rise (slower by at least one order of magnitude); iv) fast modulation ($\leq$15ns, unresolved) at turn-off followed by v) a long-lived exponential tail. We suggest a method for extracting mobility values which is found to be compatible with CW drive schemes. Mobilities for holes and electrons are extracted for a poly(p-phenylenevinylene) co-polymer and poly(di-octyl fluorene).

INTRODUCTION

The pulsed electrical excitation of polymer light emitting diodes (LEDs) has been widely reported upon [1-9] since transient electroluminescence (EL) measurements provide direct information on charge carrier processes in the device. Short duration, low duty-cycle electrical pulses minimise the Joule heating effects [6][10] often associated with device failure in constant wave (CW) conditions, and allow access to higher field regimes ($>10^7$V.cm^{-1}) [6,7]

There is, however, an absence of systematic experimental studies using electrical pulses due to the complicated interplay of injection, space charge build-up, charge transport, recombination and extrinsic heating effects in the device. Moreover, mobility values obtained using pulsed measurements have been found inconsistent with the interpretation of CW current density - voltage (j-V) measurements and / or time of flight (TOF) techniques [9]. In this paper we examine the experimental procedure for time-resolved pulsed measurements of LEDs and compare it to CW operation. We also use a phenomenological model which provides a framework for the interpretation of pulsed data.

EXPERIMENTAL

The LEDs investigated were fabricated on indium-tin oxide (ITO) coated glass substrates metallized with 100nm thick aluminium strips to reduce the resistance of the ITO sheet between the pulse generator and the active area, as reported in [11], and plasma etched to improve hole injection [12]. The active area of the diodes was 1mm^2 and the residual resistance was less than 3Ω. The polymers used in the devices were: doped polyethylene dioxythiophene/polystyrene sulphonate (PEDOT:PSS) [13,14] (used as an anode layer); a co-polymer of PPV consisting of conjugated PPV segments and non-conjugated acetylene-p-xylylene units (PPV co-polymer) [11,13] (see insert of Figure 1) and poly(di-octyl flourene) (PFO) (see Figure 3).

Two LEDs were investigated: device A (ITO/ PEDOT:PSS(50nm)/PPV co-polymer(75nm)/Ca) and device B (ITO/PFO(100nm)/Ca). The PPV copolymer film was prepared by thermal conversion (200°C for 4 hours) of the spun precursor polymer film in a dynamic vacuum of 10^{-5} mbar with an inert base-atmosphere[11]. The LEDs were electrically

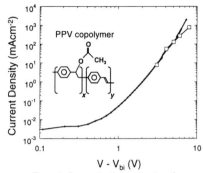

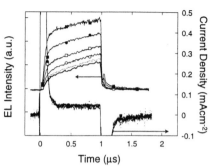

Figure 1. Current density as a function of the applied voltage minus the built-in voltage for ITO/PEDOT(50nm)/PPVco-polymer(75nm)/Ca in both CW (filled diamonds) and pulsed (empty squares) modes.

Figure 2. EL and current density response of device B to 10V 1μs pulses of varying repetition frequency. Repetition frequencies are: 100mHz (solid line); 1Hz (dashed line); 10Hz (open circles); 100Hz (open squares); 1kHz (filled circles) and 10kHz (filled square).

excited by means of a home-made pulse generator with a 10ns rise time and operating frequencies from 10mHz to 100kHz. The voltage drop across the LED was measured with a voltage probe (~10ns rise time) and the current was monitored with an induction current probe (~5ns rise time). The temporal evolution of the light output was measured using a photo-multiplier tube (PMT) with a 15ns rise time. The overall response time of the system is 15-20ns. Voltage, current and light signals were measured using a 400MHz bandwidth oscilloscope.

RESULTS AND DISCUSSION

Figure 1 shows the current density as a function of applied minus the built-in voltage for device A in both CW and pulsed mode using 10Hz repetition frequency. At low voltages there is a good agreement between pulsed and CW data, but at $j \cong 50mA.cm^{-2}$ the CW measured current starts to increase faster than the pulsed current. Numerical simulations [15] suggest that the difference in slopes and magnitudes described in Figure 1 is largely attributed to heating under CW drive conditions: the current induced heating raises the device temperature, which in turn enhances both the mobility (thermally assisted) and the current. Below 100mAcm^{-2}, however, the heating is negligible and the CW j-V curve tracks the j-V curve taken in pulsed mode. We conclude from this, therefore, that pulsed and CW j-V curves are comparable if the heating effects in CW mode are not dominant.

Transient Response of LEDs

The presence of charge storage effects in polymer LEDs must be taken into account when performing pulsed measurements, as it is important not to use pulse repetition frequencies that cause significant interference effects between successive pulses. We find that charge storage effects within a device may be easily investigated by the variation of the repetition frequency. Figure 2 shows the EL response as a function of the pulse repetition frequency, f, for device B using 10V 1μs pulses. We see that there is a threshold repetition frequency, f_0 (~100mHz), above which the integrated EL intensity of a given pulse is seen to rise. When extracting light levels or mobility values in this work, we use repetition frequencies which are sufficiently low to avoid interference effects between pulses.

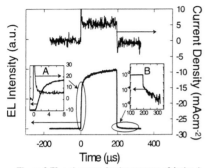

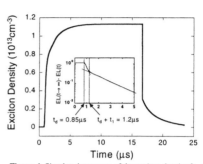

Figure 3 EL and current density response of device A to 5V 200μs pulses. Inset A shows detail of EL turn-on period and the current charging spike. Inset B shows detail of EL at turn-off on a logarithmic scale.

Figure 4. Simulated response of the exciton density for a PPV-like device (similar to device A) to a square voltage pulse (V-V_bi = 5V). Inset shows the simulated final magnitude of the EL minus the EL at time t, $EL(t \to \infty) - EL(t)$, as a function of time, showing how t_d and t_l are extracted.

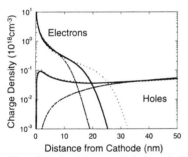

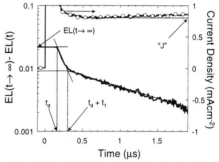

Figure 5 Simulated electron and hole charge densities as a function of the distance from the cathode density for a PPV-like device (similar to device A) for the times $t = t_d = 0.85\mu s$ (dashed line), $t = t_d + t_l = 1.2\mu s$ (solid line) and $t = 1.7\mu s$ (dashed line with large spacings).

Figure 6 Plot of the final magnitude of the EL minus the EL at time t, $EL(t \to \infty) - EL(t)$, and the current density as a function of time for device A under 10V 2μs pulses with 10Hz repetition frequency. "j" denotes the current density for the mobility calculated from SCLC theory.

Figure 3 shows the EL and current response for device A to 200μs 5V pulses. The EL time evolution is characterised by the five features: i) a delay time, t_d (~ 750ns at 5V), before light is detected; ii) a fast initial rise time, t_l (~ 4.75μs at 5V), at EL turn-on; iii) a second slower rise time, t_2 (~5ms at 5V, not shown); iv) fast turn-off modulation (of characteristic time t_3 <15ns) to a non-zero EL value followed by v) a long-lived exponential tail (of characteristic time t_4 ~~70μs at 5V). Inset A of Figure 3 shows detail of the turn-on period from which we can resolve both t_d and t_l. Inset B of Figure 3 shows detail of the turn-off period. We have found that these five transient features are also observed for devices which are injection-limited (by electrons or holes) [15].

Having established our experimental method we now describe the results of our self-consistent numerical modeling which provide a framework within which we interpret the experimental data. The model considers both drift and diffusion currents, assumes that the exciton generation is a Langevin process, and uses the field and temperature dependent mobility as given by equation 1:

$$\mu_h = \mu_{h,0} \exp\left(-\frac{\Delta}{kT}\right) \exp\left(\gamma\sqrt{E}\right) \tag{1}$$

where μ_0 is the zero-field mobility, E is the electric field, T is the device temperature and γ is a temperature dependent parameter[9]. The values used for γ, μ_0 and Δ were taken from the work done by Blom *et al.* on PPV diodes[9]. Further details of the model will be published elsewhere[15]

Figure 4 shows the response of the average exciton density to a square voltage pulse (V-V_{bi} = 5V) showing the same five features as were experimentally observed. The parameters used correspond to a ITO/PPV-like polymer/Ca device, which in the context of our model corresponds to ohmic contacts. We plot the magnitude of the EL at time t, EL(t) shown in Figure 4 subtracted from its final value, EL(t→∞), as shown in the inset of Figure 4. This plot clearly reveals the presence of two distinct regions where $\log_e$[EL(t→∞) – EL(t)] varies linearly with time. The delay time t_d is defined by the intersection of the baseline (before turn-on) and the straight line through the first slope of the $\log_e$[EL(t→∞) – EL(t)] versus time plot. Similarly $t_d + t_l$ is defined by the intersection between the straight lines drawn through the first and second slopes of the curve. t_2 is defined here as the time at which the EL intensity reaches 95% of its final.

To establish a physical picture we plot the hole and electron density adjacent to the electron injecting contact for several time delays. We chose the time delays based on Figure 4 , as t = t_d = 0.85μs, t = $t_d + t_l$ = 1.2μs and t = 1.7μs. Figure 5 shows that at time delays beyond t = $t_d + t_l$ the hole density is almost constant, with only the electron density still moving into the device. We therefore associate the second time constant (slope) in the inset of Figure 4 with the motion of the slow carriers (electrons). The use of t = $t_d + t_l$ is therefore more appropriate than using t = t_d if the deduced hole mobility is to be compared with other LED characteristics.

We now apply our mobility extraction technique to two devices made of very different polymers (PPV and PFO) (using sufficiently low repetition rates to avoid charge storage effects). Figure 6 shows the magnitude of the EL subtracted from its long-time value, EL(t→∞) – EL(t), for device A in response to 10V 2μs pulses using a 10Hz repetition frequency. The figure shows how the method used to deduce t_d and t_l from the modelled data can be applied to deduce t_d and t_l from the experimental data. Figure 7 shows the hole mobility as a function of the square root of the internal electric field calculated for the PPV copolymer by three different methods. These methods can be summarised by the following three equations:

$$\mu_1 = \frac{L^2}{(V - V_{bi})t_d} \; ; \; \mu_2 = \frac{L^2}{(V - V_{bi})(t_d + t_l)} \; ; \; \mu_3 = \frac{8}{9} \frac{jL^3}{\varepsilon_0 \varepsilon_r (V - V_{bi})^2} \tag{2-4}$$

where V_{bi} is the built-in voltage in the device (assumed to be 1.9V). μ_1 is calculated from t_d, μ_2 is calculated from $t_d + t_l$, and μ_3 is calculated entirely independently of t_d and t_l by using the current density, j, and re-arranging the SCLC equation to give the mobility. We see that the hole mobility, μ_1, result in mobility values that are approximately one order of magnitude higher than those predicted by SCLC theory (μ_3). This is in contrast to the good agreement we see between the mobility deduced using $t_d + t_l$ (μ_2) and the hole mobility, μ_3, calculated using the j-V data and equation (4) for space charge limited conduction. This agreement implies i) that the method

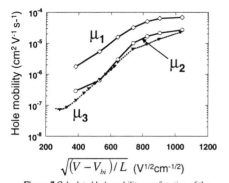

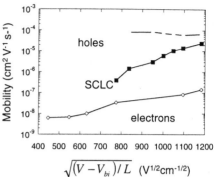

$$\sqrt{(V - V_{bi})/L} \quad (\text{V}^{1/2}\text{cm}^{-1/2})$$

$$\sqrt{(V - V_{bi})/L} \quad (\text{V}^{1/2}\text{cm}^{-1/2})$$

Figure 7 Calculated hole mobility as a function of the square root of the electric field for device A using different definitions of the carrier transit time, t_T, and the electric field, E, in the device: $t_T = t_d$, $E_{int} = V/L$ (open diamonds); $t_T = t_d + t_1$, $E_{int} = V/L$ (open circles); $t_T = t_d + t_1$, $E_{int} = 1.5$ V/L (open squares). Also shown is the mobility calculated from SCLC theory (filled triangles) from the current density vs. voltage data in pulsed mode (see Figure 1).

Figure 8 Calculated carrier mobility as a function of the square root of the electric field for device B: hole mobility calculated from $t_T = t_d + t_1$, $E_{int} = V/L$ (open circles); electron mobility calculated from $t_T = t_2$, $E_{int} = V/L$ (open diamonds) and an average carrier mobility calculated from SCLC theory (filled squares).

presented to extract the hole mobility from transient data is reliable and ii) that the device is dominated by SCLC over this field range. At fields below ~0.65MVcm⁻¹ we can describe the hole mobility by equation 1, from which we find that $\mu_{h,0} \exp\left(-\dfrac{\Delta_h}{kT}\right) = 3.5 \times 10^{-9}\, \text{cm}^{-2}\text{V}^{-1}\text{s}^{-1}$ and $\gamma_h = 1.0 \times 10^{-2}\, \text{cm}^{1/2}\text{V}^{-1/2}$. Towards 1MVcm⁻¹ the mobility starts to saturate, as has been shown in [7] for PPV (not the PPV co-polymer used here).

As a final demonstration of our method we analyse the parameters for the PFO polymer which was recently characterised using a time of flight technique [21]. As device B is hole-injection-limited we expect the density of electrons in the device to be much greater than that for holes. Because $\mu_h \gg \mu_e$ the hole mobility can therefore be extracted from t_1 data just as for device A. Due to the dominance of the low mobility electrons compared to the higher mobility holes, we therefore find that significantly less current flows in device B than in device A. Device B therefore represents an ideal system for obtaining the electron mobility from t_2 data using the following equation:

$$\mu_e = \frac{L^2}{(V - V_{bi})t_2} \tag{5}$$

where t_2 is defined here as the time at which the magnitude of the EL reaches 0.95 of its final value. Figure 8 shows the hole mobility for device B, calculated using equation (3) from the $t_1 + t_d$ data, the electron mobility using equation (5) calculated from the t_2 data, and the mobility calculated using equation (4) assuming a SCLC model. The hole mobility of PFO is found to be less field dependent in the high field regime than for the PPV co-polymer. In addition, the values calculated are found to be roughly consistent with those derived from TOF measurements [21]. The electron mobility (calculated from t_2) is found to be ~3 orders of magnitude lower than the hole mobility for a given field. We also note that the logarithm of the electron mobility increases linearly with the square root of the internal electric field as described by equation 1 from which we find that $\mu_{e,0} \exp\left(-\dfrac{\Delta_e}{kT}\right) = 7.5 \times 10^{-10}\, \text{cm}^{-2}\text{V}^{-1}\text{s}^{-1}$ and $\gamma_e = 4.4 \times 10^{-3}\, \text{cm}^{1/2}\text{V}^{-1/2}$ in good agreement with the values determined in [21]. The mobility calculated from the SCLC theory alone cannot be fitted to either the transient calculated electron or hole mobility. This is probably due to the current being composed of both electron and hole currents. In this case we may use the SCLC mobility to infer that at low fields the current is likely to be dominated by electrons, and by holes at high fields.

CONCLUSIONS

We have presented detailed analysis of pulsed excitation in polymer LEDs. Our analysis is based on universal features found in both PPV or PFO. We find that, as in the case for time of flight measurements, the method of extracting the mobility is very important. Using numerical simulations we devised a method, which is consistent both with CW drive of LEDs and probably also with time of flight technique. We have also shown that the interplay of electrons and holes needs to be accounted for, and showed that mobility values for both types of carriers can be extracted using a single device. Finally we have shown that mobility values can also be extracted for device structures that do not lend themselves to CW.

ACKNOWLEDGEMENTS

We thank the Schiff Foundation and the Engineering and Physical Sciences Research Council for financial support, and Cambridge Display Technology for supplying the polymers for this work.

REFERENCES

1   S. Karg, V. Dyakonov, M. Meier, W. Riess, and G. Paasch, Synth. Met. 67, 165-168 (1994).
2   D. Braun, D. Moses, C. Zhang, and A. J. Heeger, Appl. Phys. Lett. 61, 3092-3094 (1992).
3   H. Vestweber, R. Sander, A. Greiner, W. Heitz, R. F. Mahrt, and H. Bässler, Synth. Metals 64, PP.141-145 (1994).
4   J. Pommerehne, H. Vestweber, Y. H. Tak, and H. Bassler, Synthetic Metals 76, 67-70 (1996).
5   D. R. Baigent, P. G. May, and R. H. Friend, Synthetic Metals 76, 149-152 (1996).
6   N. Tessler, N. T. Harrison, D. S. Thomas, and R. H. Friend, Appl. Phys. Lett. 73, 732-734 (1998).
7   N. Tessler, N. T. Harrison, and F. R. H., Adv. Mater. 10, 64-68 (1998).
8   H. Chayet, R. Pogreb, and D. Davidov, Phys. Rev. B - Cond. Matt. 56, 12702-12705 (1997).
9   P. W. M. Blom and M. C. J. M. Vissenberg, Phys. Rev. Lett. 80, 3819-3822 (1998).
10   J. C. Sturm, W. Wilson, and M. Iodice, Ieee J. Selected Topics In Quantum Electronics 4, 75-82 (1998).
11   N. T. Harrison, N. Tessler, C. J. Moss, K. Pichler, and R. H. Friend, Optic. Mater. 9, 178-182 (1998).
12   J. S. Kim, M. Granstrom, R. H. Friend, N. Johansson, W. R. Salaneck, R. Daik, W. J. Feast, and F. Cacialli, Journal of Applied Physics 84, 6859-6870 (1998).
13   S. A. Carter, M. Angelopoulos, S. Karg, P. J. Brock, and J. C. Scott, Appl. Phys. Lett. 70, 2067-2069 (1997).
14   J. C. Carter, I. Grizzi, S. K. Heeks, D. J. Lacey, S. G. Latham, P. G. May, O. R. delospanos, K. Pichler, C. R. Towns, and H. F. Wittmann, Appl. Phys. Lett. 71, 34-36 (1997).
15   D. J. Pinner, R. H. Friend, and N. Tessler, (To be published).
16   M. Vanderauweraer, F. C. Deschryver, P. M. Borsenberger, and H. Bassler, Adv. Mater. 6, 199-213 (1994).
17   L. B. Schein, A. Peled, and D. Glatz, Journal of Applied Physics 66, 686-692 (1989).
18   A. Ioannidis, E. Forsythe, Y. Gao, M. W. Wu, and E. M. Conwell, Appl. Phys. Lett. 72, 3038-3040 (1998).
19   E. M. Conwell and M. W. Wu, Appl. Phys. Lett. 70, 1867-1869 (1997).
20   P. W. M. Blom and M. J. M. deJong, Ieee Journal of Selected Topics in Quantum Electronics 4, 105-112 (1998).
21   M. Redecker and D. D. C. Bradley, Appl. Phys. Lett. 73, 1565-1567 (1998).

PHOTOVOLTAGE IN CONJUGATED POLYMER PHOTOVOLTAICS WITH A TITANIUM DIOXIDE ANODE

A. C. ARANGO*, P. J. BROCK**, S. A. CARTER*
*DEPT. OF PHYSICS, UNIVERSITY OF CALIFORNIA, SANTA CRUZ, CA 95064, sacarter@cats.ucsc.edu
**IBM ALMADEN RESEARCH CENTER

ABSTRACT

We study the photovoltaic response of inorganic/organic composite devices containing the conjugated polymer, MEH-PPV, and a semi-transparent layer of titanium dioxide (TiO_2) nanoparticles, which is similar to that employed in the Grätzel photoelectrochemical cell. Comparison of the photovoltage obtainable under illumination with either a Au, Ag, Al or Ca back electrode suggests that the built in potential is determined largely by the difference between the quasi Fermi level of the TiO_2 under illumination and the work function of the back electrode. A qualitative model for charge generation and transport through the device is presented.

INTRODUCTION

The generation of photocurrent and photovoltage in a conjugated polymer based device relies on exciton dissociation and transport of excited charge to the electrodes without recombination. To improve exciton dissociation and electron transport in poly(2-methoxy,5-(2'-ethyl)-hexyloxy-p-phenylenevinylene) (MEH-PPV) devices, we introduce a porous titanium dioxide layer as the electron accepting and transporting species.[1,2,3] The TiO_2 layer, which is comprised of sintered TiO_2 nanoparticles, has been employed in Grätzel photoelectrochemical cells because of its high surface area, semi-transparency and electronic properties.[4,5]

In Grätzel cells, the maximum open circuit voltage is given by the energy difference between the quasi Fermi level of the TiO_2 under illumination and the redox potential of the replenishing electrolyte. In pure polymer photovoltaics,[6] the maximum open circuit voltage is given by the energy difference between the electrode work functions. In this paper, we examine the open circuit voltage obtainable in layered TiO_2/MEH-PPV photovoltaics with different back electrode metal contacts. By understanding the origin of the open circuit voltage, we gain insight into the charge generation and transport mechanisms that govern the overall efficiency of the device.[7]

We find that in layered TiO_2/MEH-PPV photovoltaics, the open circuit voltage depends to some extent on the difference between the TiO_2 quasi Fermi level under illumination and the workfunction of the back electrode. We conclude that the photovoltage is, to some extent, generated by the energy level difference between the back electrode and the TiO_2 layer, which serves as the anode. Additionally, a kinetically driven photovoltage may be present due to exciton dissociation at the TiO_2/MEH-PPV interface.

EXPERIMENT

TiO_2 Anatase nanoparticles in suspension (Nanoxide) were purchased from Solaronix Inc. and spread onto an ITO coated glass substrate. The substrate was sintered at 450 °C for thirty minutes. Layer thickness ranged from 1 to 5 μm. AFM images revealed nanoparticle sizes of 10-20 nm and pore sizes below 10 nm, which is most likely too small for effective penetration of MEH-PPV into the pores. MEH-PPV was spun cast onto the TiO_2 layer and was sufficiently thick to completely coat the TiO₂ surface. Au, Ag, Al or Ca electrodes were deposited on the MEH-PPV layer by thermal evaporation. We measured the current-voltage (I-V) characteristics and the open circuit voltage with a Keithley 2400 source-measure unit. A 4 Watt Xenon bulb was used for white light illumination. Samples were measured in a nitrogen atmosphere.

RESULTS

The I-V characteristics in the dark and under illumination of a pure MEH-PPV device and a device with a TiO₂ layer are compared in Figure 1. In both cases, the back electrode is Au. The dark current in forward bias in the TiO₂ layered device is suppressed due to blocking of hole injection from the ITO electrode. Under illumination in forward bias, the TiO₂ layered device photocurrent saturates at higher current values than the pure device and the zero bias short circuit current increases by almost three orders of magnitude over that of the pure device. In the TiO₂ layered device, the absence of current in the dark and the enhancement of current in the light imply photo-induced charge transfer from MEH-PPV to the TiO₂ layer, as previously identified.[1] Only a small open circuit photovoltage of almost -20 mV is observed in the pure device, while a photovoltage of -0.9 V is produced by the TiO₂ layered device.

To understand the polarity and magnitude of the open circuit voltage in the TiO₂ layered device, we construct both pure and layered devices with alternate metal contacts (Ag, Al or Ca). In Figure 2, we plot the measured open circuit voltages for devices with and without a TiO₂

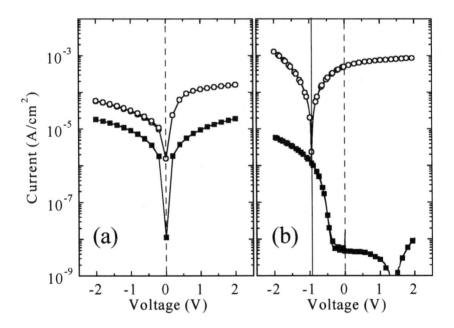

Figure 1. The I-V characteristics in the dark (open circles) and under white light illumination (filled squares) for (a) a pure ITO/MEH-PPV/Au device and (b) a layered ITO/TiO₂/MEH-PPV/Au device are plotted on a logarithmic scale. The dotted line indicates the zero bias (short circuit) state and the solid line in plot (b) indicates the photovoltage (open circuit) state.

layer versus the estimated work function of the back electrode. In addition, we plot the maximum open circuit voltage as anticipated by the metal/insulator/metal model where we consider two cases: either the ITO as anode:

$$eV_{OC} \leq eV_{bi} = \phi_C - \phi_A(ITO) \qquad (\phi_A(ITO) < \phi_C)$$

$$eV_{OC} \geq eV_{bi} = \phi_C - \phi_A(ITO) \qquad (\phi_A(ITO) > \phi_C)$$

or the TiO$_2$ as anode:

$$eV_{OC} \leq \phi_C - \varepsilon_F(TiO_2) \qquad (\varepsilon_F(TiO_2) < \phi_C)$$

$$eV_{OC} \geq \phi_C - \varepsilon_F(TiO_2) \qquad (\varepsilon_F(TiO_2) > \phi_C)$$

where V_{OC} is the open circuit voltage, V_{bi} is the built in voltage, $\phi_A(ITO)$ is the work function of the ITO, ϕ_C is the work function of the back contact or cathode and $\varepsilon_F(TiO_2)$ is the quasi Fermi level of the TiO$_2$ layer under illumination. $\phi_A(ITO)$ is assumed to be -4.9 eV and $\varepsilon_F(TiO_2)$ is assumed to be -4.0 eV.

The measured values of V_{OC} for pure devices roughly correspond to the expected values for the ITO anode case, yet are lower than predicted because of interfacial charge injection at

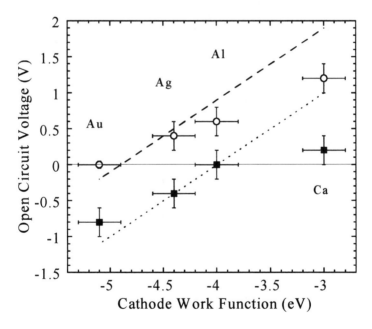

Figure 2. The measured open circuit voltages for pure MEH-PPV devices (open circles) and layered TiO₂/MEH-PPV devices (closed squares) are plotted versus the approximate work function of the back electrode or cathode. The difference between the ITO anode and the cathode work functions is plotted with a dashed line and the difference between the quasi Fermi level of the TiO₂ under illumination and the cathode work function is plotted with a dotted line.

room temperature.[8] In contrast, V_{OC} of the TiO$_2$ layered devices appear to coincide with the expected values for the TiO$_2$ anode case although V_{OC} for Ca is significantly lower than expected while V_{OC} for Ag is greater (in magnitude) than expected.

We can understand the observed downward shift in V_{OC} in TiO$_2$ layered devices versus pure devices and the slight deviation from the metal/insulator/metal model by considering photo-injection of electrons into TiO$_2$ from MEH-PPV. Photo-induced electron transfer from MEH-PPV to TiO$_2$ will result in population of empty states in the TiO$_2$ conduction band and in elevation of the TiO$_2$ Fermi level.[5] According to the condition for thermal equilibrium, the Fermi levels of TiO$_2$ and ITO will align if particles can move freely between these two materials. The discontinuity between this Fermi level and that of the cathode metal is the expected photovoltage. However, a kinetically driven photovoltage, which arises from the discontinuity between the conduction band of MEH-PPV and the conduction band of TiO$_2$, may also be present. Such a photovoltage can either compete (Ca and Al cathode devices) or compliment (Au and Ag cathode devices) the internal field generated by electrode asymmetry.[9]

The energy level diagram of the TiO$_2$ device under illumination and in short circuit conditions (Figure 3) helps illustrate the charge generation, injection and transport processes mentioned above. An incident photon in the visible passes through the TiO$_2$ layer and is absorbed by the polymer whereby an exciton is produced and may diffuse a distance of up to 10-20 nm towards the TiO$_2$/polymer interface.[2] Interfacial exciton dissociation can occur via electron transfer to the TiO$_2$ - resulting elevation of the TiO$_2$ Fermi level. Electron diffusion through the TiO$_2$ layer,[10] injection into the ITO and redistribution of charge through the external circuit such that the TiO$_2$, ITO and Au Fermi levels are aligned produces an internal field across the polymer layer. Note that we do not consider band bending or other interfacial barrier effects.

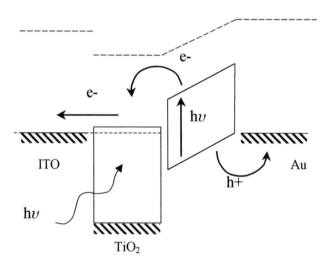

Figure 3. The energy level diagram and charge transfer processes under short circuit conditions and under illumination for a layered ITO/TiO$_2$/MEH-PPV/Au device.

CONCLUSION

We have measured the open circuit voltage in polymer photovoltaics where a TiO_2 layer is inserted between the ITO and the polymer layer. We find that the open circuit voltage depends on the work function of the cathode in a way that is roughly consistent with the metal/insulator/metal model when the TiO_2 layer is considered to be the anode. We speculate that exciton dissociation at the TiO_2/MEH-PPV interface produces an additional photovoltage. We suggest that charge generation at zero bias involves exciton dissociation at the TiO_2/MEH-PPV interface, diffusive charge transport through the TiO_2 layer and an internal field across the polymer layer.

REFERENCES

[1] A. C. Arango, P. J. Brock, and S. A. Carter, Appl. Phys. Lett. **74** (12), p. 1698-1700 (1999).

[2] T. J. Savenje, J. M. Warman, and A. Goossens, Chem. Phys. Lett. **290**, p. 297-303 (1998).

[3] T. K. Däubler, I. Glowacki, U. Scherf, J. Ulanski, H.-H. Hörhold, D. Neher, submitted to J. Appl. Phys.

[4] B. O'Regan, and M. Grätzel, Nature **353**, p. 737-9 (1991).

[5] A. Hagfeldt, and M. Grätzel, Chem. Rev. **95**, p. 49-68 (1995).

[6] R. N. Marks, J. J. M. Halls, D. D. C. Bradley, R. H. Friend, and A. B. Holmes, J. Phys.: Condens. Matter **6**, p. 1379-94 (1994).

[7] K. Yoshino, K. Tada, A. Fuji, E. M. Conwell, and A. A. Zakhidov, IEEE Trans. Elec. Dev. **44** (8), p. 1315-1323 (1997).

[8] G. G. Malliaras, J. R. Salem, P. J. Brock, and J. C. Scott, J. Appl. Phys. **84** (3), p. 1583-7 (1998).

[9] B. Gregg, Appl. Phys. Lett. **67** (9), p. 1271-3 (1995).

[10] A. Solbran, H Lindstrom, H. Rensmo, A. Hagfeldt, S. Linquist, and S. Sodergren, J. Phys. Chem. B **101**, p. 2514-8 (1997).

TOWARDS THE SYNTHESIS OF MONO- AND BI-FUNCTIONAL AMPHIPHILIC OLIGOTHIOPHENES: ORGANIC MATERIALS FOR OPTO-ELECTRONIC APPLICATIONS

R.C. ADVINCULA[1], S. INAOKA[1], M. PARK[1,2], D. PHILLIPS[1], D.M. SHIN[2]
[1]Department of Chemistry, University of Alabama at Birmingham, Chemistry Building, Birmingham, AL 35294-1240, USA
[2]Department of Chemical Engineering, Hong-Ik University, Seoul, Korea

ABSTRACT

In this report, we describe our initial synthesis and characterization of mono-functional and bi-functional dibromoalkyl oligothiophenes to achieve amphiphilicity and telechelic functionality. Oligothiophenes are an important class of organic materials for opto-electronic devices and display applications. We have mono-functionalized oligothiophenes by the synthesis of a quinquethiophene bromoalkyl derivative. A bi-functional sexithiophene was derived primarily by the symmetrical coupling of terthiophene derivatives. Both were synthesized using Grignard coupling and lithiation reaction methodologies. UV-Vis, IR, NMR, MALDI-TOF-MS, and DSC confirmed the structure and physical properties of the oligomers. In addition, we have also synthesized an amphiphilic diamine derivative from the reaction of hexamethylenediamine with a bromoalkyl terthiophene derivative. Using photoluminescence, the photophysical properties of the oligomers were found to be that of typical oligothiophenes. Processing as ultrathin films for devices is currently being investigated.

INTRODUCTION

Oligothiophenes with well-defined structures such as α-sexithiophene continue to attract much attention. They have well-ordered structures,[1] which lead to good charge transport properties and high carrier mobility.[2] Its alkyl group substituted analogues have a large anisotropy of conductivity and high field-effect carrier mobility because of inherent self-organizing trends.[3] Thus, these materials have the potential for use in opto-electronic thin-film devices including field-effect transistors, FET,[4] light-emitting devices, LEDs,[5] and Schottky-type diodes.[6]

The design of organic and polymer materials for opto-electronic applications requires a combination of functionality and processability. For ultrathin film fabrication techniques using wet processes, e.g. spin-coating, molecular assembly, etc., the materials have to be soluble to organic or aqueous solvents or amphiphilic. While vacuum deposition is a technique commonly applied to small molecules and oligomer derivatives, ultrathin film geometries for devices can also be realized by self-assembly approaches.[7] These molecular and macromolecular self-assembly approaches, e.g., self-assembled monolayers (SAM), Langmuir-Blodgett (LB) films, alternate polyelectrolyte deposition (APD), spin-coating, etc. have been repeatedly applied in fabrication of devices made of organic solid-state materials.[8] The novelty of such approaches is in providing an alternative to crystalline packing structures for fundamental or applied investigations and the possible advantage of continuous batch processing.

To address this possibility, we have attempted to functionalize oligothiophenes toward amphiphilic properties. In this work, we report the synthesis of oligothiophenes (terthiophene, quinquethiophene, sexithiophene) and characterized their initial photoluminescence properties.

155

The oligomers were synthesized by a combination of Grignard coupling and lithiation reactions. In addition, we have synthesized a diamine derivative from the reaction of *N, N*-dimethyl-1, 6-hexanediamine to form *N,N'*-dimethyl-*N,N'*-bis(6-(α-terthienyl)hexyl)-1,6-hexanediamine. These materials can eventually be functionalized as an amphiphile or telechelic oligomer to form ionene polymers with oligothiophene backbone or side groups.[9]

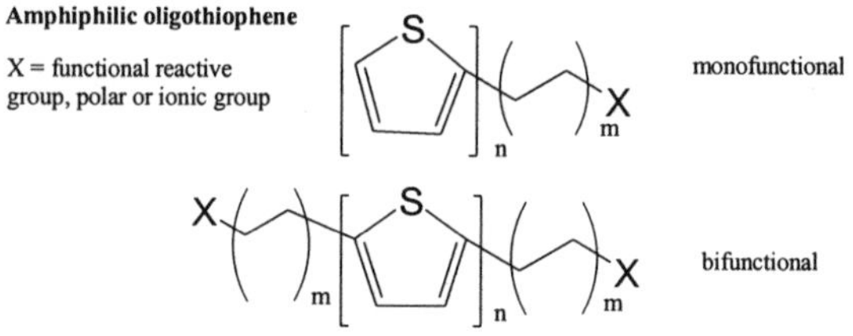

Figure 1. Schematic representation of mono- and bi-functional oligothiophene amphiphiles.

Amphiphilic oligothiophene

X = functional reactive group, polar or ionic group

monofunctional

bifunctional

EXPERIMENT

Materials. All starting reagents and compounds of sufficient purity with the exception of freshly distilled solvents (THF and DMF) were used as received (Aldrich Chemical Co.). 2,2'-bithiophene is a commercially available product (Aldrich) and was purified by silica gel column chromatography (hexane).

Instrumentation. NMR (Bruker 400 MHz), IR (Bruker Vector 22), UV-vis (Perkin Elmer Lambda 20), MALDI-TOF-MS (Perseptive Biosystems Voyager Elite DE), TGA-DSC (Mettler TG50, DSC30), PL (Perkin Elmer LS50B). For monolayer investigations, the KSV LB 2000 Dipping trough was used under a laminar flow hood conditions. Surface pressure area isotherms, isobaric creep measurements, and compression-expansion cycles (hysteresis) were investigated at ambient temperatures (25 °C).

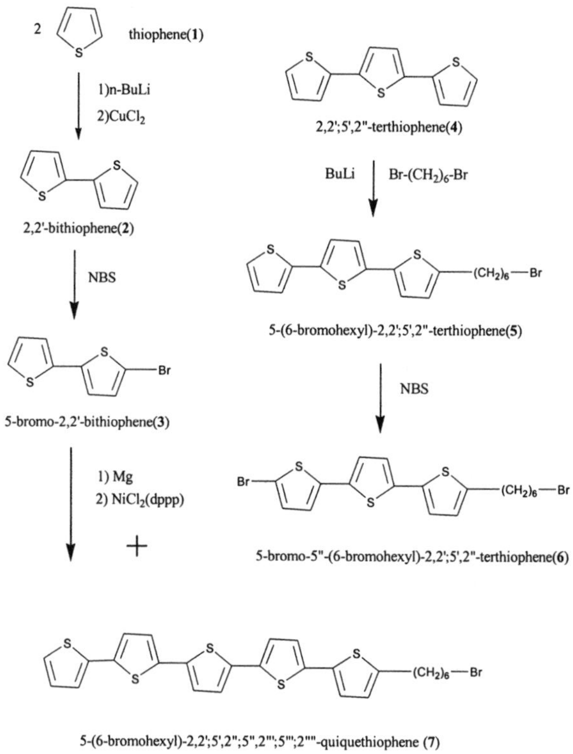

Figure 2. Synthesis scheme for 5-(6-bromohexyl)2,2';5',2";5",2"';5"',2""-quinquethiophene.

A full description of the synthesis procedure and characterization for the derivatives in each step is the subject of a forthcoming publication.[10] The following are a description of some spectral properties of each oligothiophene product:

Synthesis of 5-(6-bromohexyl)-2,2';5',2";5",2'";5'",2""-quinquethiophene. The properties of the product (7) are summarized as follows: ^{1}H-NMR(CDCl$_3$), δ(6.98, m, 4H), δ(6.89, d, 1H), δ(6.69, d, 1H), δ(3.41, t, 2H), δ(2.80, t, 2H), δ(1.85, p, 2H), δ(1.68, p, 2H), δ(1.42, p, 4H) and UV-vis in CHCl$_3$, λ_{max}=416 nm., MALDI-TOF-MS confirmed the MW = 574, TGA showed a decomposition onset at 275 °C, IR spectroscopy reveal the presence of the corresponding CH$_2$ stretching frequency of the alkyl chain (3070 cm^{-1}), CH$_2$-Br stretching (1254 cm^{-1}) and substituted thiophene frequencies (834-695 cm^{-1}).

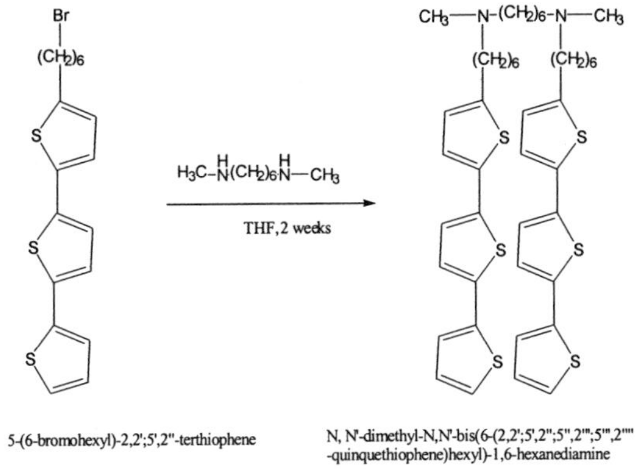

Figure 3. Scheme for the synthesis of N,N'-dimethyl-N,N'-bis (6-(2,2';5',2"; 5",2'";5'",2""-terthienyl) hexyl)-1,6-hexanediamine.

Synthesis of N,N'-dimethyl-N,N'-bis (6-(2,2';5',2";5",2'";5'", 2"" -terthiophene) hexyl)-1,6-hexanediamine. This product was characterized by ^{1}H-NMR(CDCl$_3$), δ(7.18, dd, 2H), δ(7.12, dd, 2H), δ(6.97, m, 8H), δ(6.94, d,2H), δ(2.77, m, 12H), δ(2.62, s, 6H), δ(1.87, m, 8H), δ(1.64, p, 4H), δ(1.37, m, 12H) and UV-vis in CHCl$_3$, λ_{max} = 364nm.

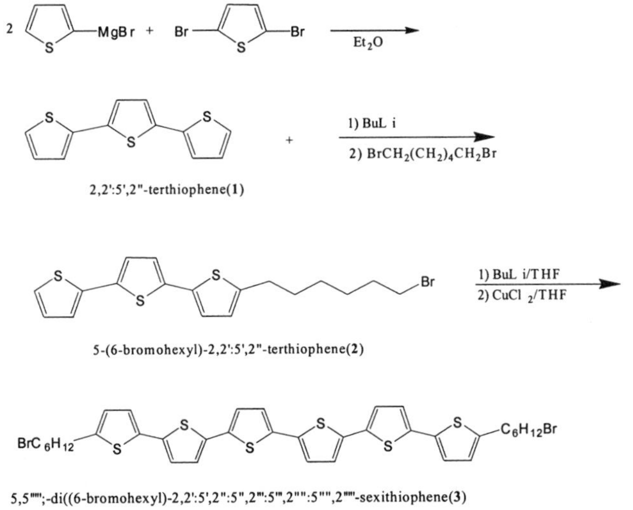

Figure 4. Scheme for the synthesis of 5,5""''-di((6-bromohexyl)- 2,2': 5',2": 5",2"': 5"', 2"":5""',2"""'-sexithiophene.

Synthesis of 5,5""''-di((6-bromohexyl) 2,2': 5',2": 5",2"': 5"', 2"":5""',2"""'-sexithiophene. The product was only slightly soluble in ordinary organic solvents, which inhibited NMR analysis. It has a reddish color and by UV-VIS analysis, showed visible absorption at the 400-500 nm region. IR spectroscopy reveal the presence of the corresponding CH$_2$ stretching frequency of the alkyl chain (3070 cm^{-1}) as well as CH$_2$-Br stretching (1254 cm^{-1}) and substituted thiophene frequencies (834-695 cm^{-1}). MALDI-TOF-MS confirmed the MW of the product = 784 g/mol. TGA analysis reveal an onset of decomposition at 323 °C. We are currently purifying the product further using several recrystallization and sublimation procedure (primarily by the method of Katz and co-workers)[12] prior to obtaining m.p. data. X-ray diffraction will be done to elucidate degree of crystallinity and ordering.

RESULTS AND DISCUSSION

Although the synthesis of oligothiophene derivatives have been reported almost 50 years ago, the importance of purity and yield needs to be addressed since applications for electro-optical devices demand these standards.[11] Recently a methodology reported by Katz et. al. outlined the synthesis of dopant-free thiophene hexamers.[12] Their methodology made use of ferric acetylacetonate as the organolithium reagent.

Synthesis of the monofunctional derivative was done primarily using the asymmetrical coupling route of a terthiophene and bithiophene derivatives with a Grignard reagent. We attempted to synthesize a sexithiophene derivative with 5-bromo-2,2':5',2"-terthiophene derivative using a Grignard reagent for coupling with 5-bromo-5"-(6-bromohexyl)-2,2';5',2"-terthiophene but had little success. We attributed this to a drastic decrease in solubility with common organic solvents and difficulty in purification. The asymmetric coupling was best achieved by functionalizing the α-3T prior to reacting with the bithiophene. One advantage of these scheme being the effective separation between the bromohexylterthiophene derivative and the unreacted terthiophene and bromohexane by simple filtration and silica gel column chromatography. The final product (7) of Figure 2, gave a reddish color due to the more extensive π-conjugation along the molecule long axis. The UV-VIS analysis showed a significant shift in absorption maximum, with the bandgap energy lowered with the appearance of a peak at the 416 nm and a shoulder at the 600-700 nm region. A Photoluminescence study of (7) (from Figure2), showed the predictable result of higher luminescence with increasing excitation wavelength and energy. The best result came at an excitation wavelength of 420 nm, which resulted in luminescence observed which peaked at approximately 520 nm.

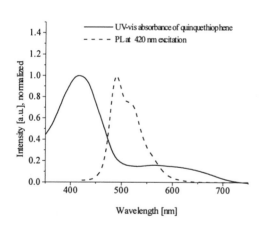

Figure 5. UV-vis and PL (420 nm, excitation) for Product (7).

For the bifunctional oligothiophene, the synthesis was comparable to the scheme previously reported by Michalitsch et.al. using symmetrical coupling.[13] In our methodology, the symmetric coupling was best achieved by functionalizing the α-3T prior to dimerization. In general, a minimum of purification techniques was necessary for each step, primarily using column chromatography, followed by recrystalization in the second, and continuous extraction in the third step. Product (3) as shown in Figure 4, had limited solubility in ordinary organic solvents, inhibiting us from doing NMR analysis and requiring long continuous extraction. It had a reddish color due to the more extensive π-conjugation along the molecule long axis. This was confirmed in the UV-VIS analysis of product (3) which showed visible absorption at the 400-500 nm region corresponding to the sexylthienyl derivative as described in literature. The peak at the 361 nm region did not completely disappear, indicating the presence of some terthiophene material on the crude product. IR, MALDI-TOF, DSC, confirmed the structural and physical properties of the synthesized product. We are currently purifying samples for X-ray diffraction analysis.

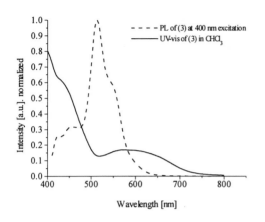

Figure 6. UV-vis and Photo-luminescence data for product (3).

A Photoluminescence study of (3) showed the predictable result of higher luminescence with increasing excitation wavelength and energy. The best result came at an excitation wavelength of 400 nm, which resulted in luminescence observed at approximately 520 nm.

We also investigated the amphiphilic nature of the diamine derivative. It showed ideal surface-pressure area isotherm behavior as monolayer at the air-water interface (compressed at various rates from 3.0-7.5 Å^2 /molecule x min) Equilibrium isotherm measurements will be done towards lower compression rates. The mean molecular area (MmA), corresponded to more than twice that of the ideal area for one repeat unit incorporating the oligothiophene (55 Å^2 /molecule). The Π-A area isotherms is shown in Figure 7. Careful analysis of their aggregation behavior, tilt orientation of the chromophoric groups, and compressibility, will give us further explanation about these high values. Repeated compression-expansion cycles (hysteresis) showed *reversible* behavior at a compression-expansion rate up to 7.5 Å^2 /molecule x min. Isobaric creep measurements showed monolayer film stability up to 1 hour (area decrease of less than 1 Å^2/molecule). We therefore expect these films to be suitable for further studies in multilayer LBK depositions (vertical).

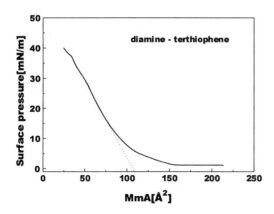

Figure 7. Surface Pressure-Area isotherm of the diamine amphiphile at 25 °C (7.5 Å^2 /molecule x min) with reproducibility within 1 Å^2 /molecule.

CONCLUSION

We have synthesized several functional oligothiophene derivatives. The synthesis and analysis of these oligomers are straightforward, with the length of the oligomer limited by solubility. The photoluminescence properties showed typical oligothiophene photophysical properties, the difference being the functionality for further synthesis as an amphiphile or telechelic. The amphiphilicity was demonstrated with the diamine derivative in Langmuir-monolayer experiments. Further studies will be done to explore their telechelic functionality in the synthesis of ionene polymers.

ACKNOWLEDGMENT

Faculty Research Grant program of UAB, SURE Program for International students, and the Department of Chemistry start-up grant.

REFERENCES

[1] A. Lovinger, D. Davis, A. Dodabalapur, H. Katz, Chem. Mater. **8**, 2836 (1996).
[2] B. Servet, G. Horowitz, S. Ries, O. Lagorsse, P. Alnot, A.Yassar, F. Deloffre, P. Srivastava, R. Hajlaoui, P. Lang, F. Garnier, Chem. Mater. **6**, 1809 (1994).
[3] F. Garnier, A.Yassar, R. Hajlaoui, G. Horowitz, F. Deloffre, B. Servet, S. Ries, P. Alnot, J. Am. Chem. Soc. **115**, 8716 (1993).
[4] G. Horowitz, F.Garnier, A. Yassar, R. Hajlaoui, F. Kouki Adv. Mater. **8**, 52 (1996).
[5] F. Cacialli, R. Marks, R. Friend, R. Zamboni, C. Taliani, S. Maratti, A. Holmes, Synth. Met. **76**, 145 (1996).
[6] D. de Leeuw, E.J. Lous, Synth. Met. **65**, 45 (1994).
[7] A. Ulman, *Ultrathin Organic Films*, (Academic Press: San Diego, 1991).
[8] *Electrical, Optical, and Magnetic Properties of Organic Solid-State Materials IV*. Eds. J. Reynolds, A.K. Jen, M. Rubner, L. Chiang and L. Dalton, MRS Proceedings, (MRS, Pennsylvania, Vol. 488, 1998).
[9] K. Mullen, J. Lex, R. Schulz, F. Wakter, Polymer Bull. **24**, 263 (1990); G. Hochberg, R. Schultz, Polymer Intl. **32**, 309 (1993).
[10] M. Park, D. Shin, S. Inaoka, R. Advincula, manuscript in preparation.
[11] J. Sease, and L. Zechmeister, J. Am. Chem. Soc. **69**, 270 (1947).
[12] H. Katz, A. Dodabalapur, L.Torsi, A. Lovinger, and R. Ruel, Polym . Mat. Sci. Eng. Preprints, **72**, 467 (1995).
[13] R. Michalitsch, P. Lang, A. Yassar, G. Nauer, F. Garnier, Adv. Mater. **9**, 321 (1997).

DETERMINATION OF CRYSTALLITE SIZE AND LATTICE STRAIN IN HEXAPHENYL THIN FILMS BY LINE PROFILE ANALYSIS

H.-J. BRANDT, R. RESEL, J. KECKES*, B. KOPPELHUBER-BITSCHNAU**, N. KOCH AND G. LEISING
Institute for Solid State Physics, Technical University Graz, AUSTRIA
*Department of Theoretical Chemistry, Institute of Inorganic Chemistry,
 Slovak Academy of Sciences, Bratislava, SLOVAKIA and
 Erich Schmid Institute, Austrian Academy of Sciences, Leoben, AUSTRIA
**Institute for Physical and Theoretical Chemistry, Technical University Graz, AUSTRIA

ABSTRACT

Hexaphenyl thin films (HTF) are widely used as an electro-active organic medium in blue light emitting diodes. The optical parameters of the HTF-based devices significantly depend on their microstructural properties.

HTF of different types are produced by physical vapor deposition on glass substrates applying specific sample preparation conditions. The microstructural properties of HTF are characterized using X-ray diffraction line profile analysis and atomic force microscopy (AFM). Diffraction peaks representing three different types of preferred growth in HTF are analyzed, namely textures with (00λ), $(22\bar{3})$ and $(20\bar{3})$ net planes oriented parallel to the substrate. No additional line-broadening (compared to silicon powder used as standard) is observed in the case of a film prepared at high substrate temperature of 170 °C. On the other hand, considerable broadening is detected in a film with the substrate kept at room temperature. Multiple line analysis documents that the crystallite size and lattice strain for the sample is 150 nm and 3×10^{-4}, respectively. Single line analysis performed on the other reflections reveal size-induced broadening for a crystallite size in the range 40 to 50 nm. From AFM data we obtained that the maximum roughness of the surface is about 40 nm. The results indicate that the deposition temperature significantly influences the microstructural properties and that higher substrate temperature promotes a higher mobility of the molecules on the substrate enabling growth of larger crystallites with lower strain.

INTRODUCTION

Hexaphenyl is an electro-active organic material with high potential for future applications in light emitting devices [1,2]. By depositing the molecules on isotropic indium tin oxide (ITO) substrates, it was demonstrated that the HTF can be successfully applied in color flat panel displays [3]. Since the optical as well as the electrical properties of HTF are highly anisotropic detailed structural information are necessary for good performance in the applications [4]. The crystal structure of Hexaphenyl, obtained from single crystal analysis, is monoclinic with the lattice constants of a=8.091, b=5.568, c=26.24, ß=98.17° and the space group P21/a [5]. Using X-ray diffraction it was demonstrated that it is possible to prepare HTF of crystalline form [6]. Three different types of preferred orientation of the crystallites were observed in the films. Namely, two independent and relatively strong fiber textures characterized with i) the (001) planes and ii) the $(22\bar{3})$ plane parallel to the surface were detected [7,8]. The third type of preferred orientation was a bi-axial texture obtained by a rubbing procedure of the HTF [9]. In this case the $(20\bar{3})$ plane was found parallel to the substrate.

The optical measurements as well as theoretical studies demonstrated that the optical properties of the HTF depends significantly on the preferred orientation [10]. Moreover, optical experiments with polarized light revealed that also the microstructure, namely crystallite size and strain, influences the internal emission/absorption processes in HTF [11].

Line profile analysis is a powerful method to obtain reliable microstructural characteristics. Usually applied to thin films of inorganic materials like high Tc superconductors, Indium Tin Oxide (ITO) or coating materials [12,13,14,15], line profile analysis on thin organic films is a difficult task. The problem is that diffraction peaks of sufficient intensity can be observed only at high d-values and, moreover, no standards for the determination of the instrumental broadening is available in this region. Another problem is the high sample transparency of organic materials.

EXPERIMENTAL DETAILS

Film Preparation

Hexaphenyl of high purity was purchased from Tokyo Chemical Industries Ltd. Thin films were prepared by physical vapor deposition of Hexaphenyl molecules on isotropic glass substrates in high vacuum (4.10^{-4} Pa). The temperature of the quartz crucible was set to 275°C, a deposition rate of Hexaphenyl on the substrate of 0.1 Å/sec was obtained. The first layer was deposited with a thickness of 9 nm by keeping the substrate at room temperature. This layer was rubbed over a tissue and the final thickness of the film was obtained by continuing the Hexaphenyl deposition. With this sample preparation conditions and with this special rubbing technique [9] we obtained different preferred orientations of the crystallites. Two different types of films were prepared. The substrate temperature was varied, in one case ambient temperature was chosen (room temperature (RT) sample) and in the second type the substrate temperature was 170°C (high temperature (HT) sample) [16]. The deposition rate and thickness of the samples were controlled by a quartz-microbalance and checked by an atomic force microscope operated in the tapping mode. The total thickness of the HT-sample was 4.1 µm, the thickness of the RT-sample 4.4 µm.

X-ray diffraction measurements

The diffraction pattern was collected with a Philips PW 1820 diffractometer in Bragg-Brentano geometry. Only reflections of net planes parallel to the surface of the substrate are detected in this geometry. Line profile analysis of reflections in this configuration reveals only size & strain information perpendicular to the substrate surface. Copper $K\alpha$ radiation and a secondary monochromator was used. The patterns were recorded between 3 and 43 degrees with a step size of 0.02 degree, the peaks of strongest intensity of Hexaphenyl are located at this low diffraction angles. Scan time was 80 s per step except of the two 001 reflections with just 10 s scan time. This long scan times are important because of the low intensities for thin organic films. A high background arises due to the glass substrates. The measurement was carried out under room temperature and atmospheric pressure. Diffraction peaks of a high purity silicon powder (National Bureau of Standards, Standard Reference Material 640b) with a mean particle size of about 5µm were taken for comparison of line breadth.

Data evaluation

The measured peaks of both samples were fitted with STOE Powder Diffraction Software. Symmetrical Pseudo-Voigt function was chosen including a correction for the $K\alpha_2$ contribution. The obtained parameters of full width at half maximum ($2w$) and integral breadth (β) were considered for further evaluation of microstructural properties. Multiple line analysis of β was done in the case of (00λ) reflections. They are analyzed in a Williamson-Hall-plot [17] using the relationship

$$\beta \cdot \cos\theta = \frac{\lambda}{D} + 4 \cdot e \cdot \sin\theta$$

where θ is the half scattering angle, β is the integral breadth of the reflections in radians, λ is the wavelength of the incident beam, D is the apparent crystallite size and $e=\Delta d/d$ is the microstrain.

The single line Voigt analysis was carried out in the case of other reflections (11$\bar{1}$,11$\bar{2}$ and 20$\bar{3}$) of the HTF. From the resulting values of (2w) and β we calculate Cauchy and Gaussian contribution of the line profile. To get the corrected integral breadth we separate the Lorentz and Gauss contribution of the broadened peaks and subtract the respective part of broadening of the standard. From broadening relative to the standard we obtain average size from the Cauchy part of broadening (β^f_C) and strain from the Gauss part (β^f_G) [18,19].

$$D = \frac{\lambda}{\beta^f_C \cos\theta} \quad \text{and} \quad e = \frac{\beta^f_G}{4\tan\theta}$$

RESULTS AND DISCUSSION

The diffraction pattern of two Hexaphenyl samples are shown in Figure 1 with intensities shifted relative to each other and plotted in a logarithmic scale. The strong progression of the background arises due to the scattering on the amorphous glass substrate. The 001 peaks was measured seperatly with a shorter integration time and showed therefore this low intensities in Figure 1. Peaks representing three different preferred orientations of the crystallites are observed. The first orientation is represented by 001, 002, 003, 004, 005, 006 and 00$\underline{12}$ and this type of preferred orientation is characterized by the (00λ) plane parallel to the surface of the substrate [7]. The second orientation, the (22$\bar{3}$) plane parallel to the substrate, has the two reflections 11$\bar{1}$ and 11$\bar{2}$ [8]. The third orientation with (20$\bar{3}$) parallel to the substrate shows only the 20$\bar{3}$ peak [9].

In Figure 1 we see a double peak structure quite strong in the case of 001, 002 and 003 peaks of the RT-sample, as well as in the 001 and 002 peaks of the HT-sample due to another

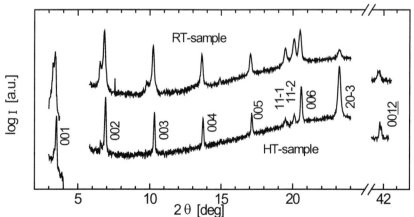

Figure 1: θ / 2θ scan of the RT and HT sample, intensities are plotted in a logarithmic scale and shifted relative to each other

polymorph phase of Hexaphenyl [6]. They are considered by the profile fitting procedure but not further evaluated. Because of the overlap of both polymorph phases on the 001 reflections a definite determination of the integral breadth was not highly accurate. In general, the fitting procedure was simplified by a small overlap of the peaks, only three peaks (11 $\bar{1}$, 11 $\bar{2}$ and 006) are at close quarters and had to be fitted together. The 00$\underline{12}$ reflection is despite its low intensity of 200 counts above the background suited for fitting as peak at high 2θ angles. In a general overview of Figure 1 we see that the peaks of the RT-sample are significantly more broadened than the peaks of the HT-sample. This suggests either smaller crystallite size or higher strain in the crystal lattice of RT sample.

Figure 2 shows an example of the measured data of the HT-sample and the resulting fitted curve in a range around 20 degrees with the 11 $\bar{1}$, 11 $\bar{2}$ and 006 peaks. The intensity values are given in linear scale and show the real measured data. Symbols are measured data, the full line represents the fit, the dashed line is the background.

The determined integral breadth of all observed reflections of the RT-samples (squares) and HT-samples (circles) and also the integral breadth of the silicon standard peaks (triangles) were plotted in Figure 3. Silicon has reflections only above 2θ = 28°. The integral breadth of the silicon reflections were linear fitted and extrapolated to get values of low diffraction angles. We obtain good agreement of this fit with the results of the (00λ) plane reflections of the HT-sample (full circles). We take the fit of the 00λ peaks of the HT-sample as standard at lower 2θ angles. At basis of this new standard we analyze the RT-sample.

The (00λ) reflections of the RT-sample (filled squares) show a significant broadening relative to the HT-sample. The peaks for crystallites of the (22 $\bar{3}$) orientation (11 $\bar{1}$ and 11 $\bar{2}$ reflections) and of the (20 $\bar{3}$) orientation (20 $\bar{3}$ reflection) (hollow squares) were more broadened than the 00λ reflections and therefore have other microstructural properties. Also reflections of the HT-sample of these orientations (hollow circles) were broadened relative to the 00λ reflections.

A multi line analysis of line breadth for evaluation of size and strain of the 00λ reflections of the RT-sample is done by the Williamson-Hall plot (Figure 4). For this task we subtracted the standard (linear fit of 00λ reflections of the HT-sample) from the integral breadth of the RT peaks. From the linear fit we obtain 151 nm for crystallite size and for lattice strain 3×10^{-4}, but as mentioned in literature the value for size might be overestimated [17,19].

For the peaks of other orientations we estimate size and strain with the single line Voigt analysis [18]. We determine a size of 52 nm for 11 $\bar{1}$, 48 nm for 11 $\bar{2}$ – this gives an median value of 50 nm for the (22 $\bar{3}$) orientation - and 40 nm for (20 $\bar{3}$). For comparison we did the single line analysis also with the (00λ) reflections. Here we get values for size from 102 nm to 148 nm. The median value for that is 118 nm, which is slightly smaller than our results of the Williamson Hall plot. But we see, that the size of (00λ) orientations are about double in size than crystallites of the other orientations.

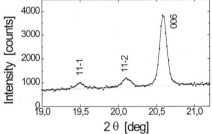

Figure 2: Data fitting of peak 11-1, 11-2 and 006 of the HT-sample. The dashed line is the background.

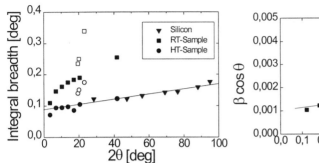

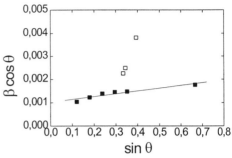

Figure 3: Integral breadth of the RT, HT and silicon diffraction peaks. The black line is an fit through the integral breadths of silicon. Open Symbols mark different orientations.

Figure 4: Williamson Hall plot for the RT-sample, black line is an fit through the peaks 002 to 006 and 0012. Open Symbols mark different orientations.

The AFM pictures are depicted in Figure 5, on the left side we see the RT-sample and the HT-sample on right. The micrographs of 4 µm to 4 µm shows the surface of the samples and below an example of the section analysis. From AFM data we obtained that the maximum roughness of the RT-surface is about 40 nm, roughness of the HT-face is in average 100 nm. In comparison to the results of the line breadth analysis the roughness is smaller than crystallite size. The section analysis gives also proof, that lateral size of the crystallites of the HT-sample are bigger than that of the RT-sample.

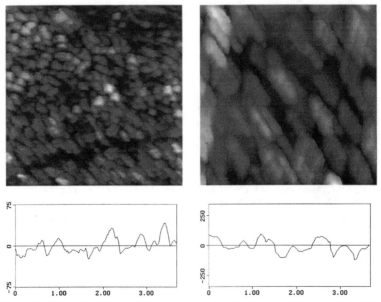

Figure 5: AFM micrograph of 4 x 4 µm (above) and section analysis (below) of the RT-sample (left) and the HT-sample (right).

CONCLUSION

Different preferred orientations of the crystallites of HTF result in their individual line broadening and consequently different microstructural properties. We exhibit, that the contribution of broadening is mainly determined by the crystallite size. We could show, that for high substrate temperatures the crystallites grows in much larger size due to a higher mobility of the molecules on the surface of the substrate. We obtained crystallite sizes greater than 150 nm for the (00λ) orientation, 50 nm for the $(22\bar{3})$ and 40 nm for the $(20\bar{3})$ orientations.

ACKNOWLEDGEMENT

We thank the *Aktion Österreich-Slowakei* under project no. 22s29 for support of this work, further the Industriellenvereinigung Steiermark, Land Oberösterreich, the Österreichische Forschungsgemeinschaft and the Rektor der Technischen Universität Graz for financial support.

REFERENCES

1. W. Graupner, G. Grem, F. Meghdadi, Ch. Paar, G. Leising, U. Scherf, K. Müllen, W. Fischer and F. Stelzer, *Mol. Cryst. Liq. Cryst.* **256**, 549 (1994).
2. G. Leising, S. Tasch and W. Graupner in *Handbook of Conducting Polymers*, edited by T.A. Skotheim, R.L. Elsenbaumer and J.R. Reynolds (Marcel Dekker, Inc., New York, 1998), p. 847.
3. S. Tasch, C. Brandstätter, F. Meghdadi, G. Leising, G. Froyer and L. Athouël, *Adv. Mater.* **9**, 33 (1997).
4. H. Yanagi and S. Okamoto, *Appl. Phys. Lett.* **71**, 2563 (1997).
5. K.N. Baker, A.V. Fratini, T. Resch, H.C. Knachel, W.W. Adams, E.P. Socci and B.L. Farmer, *Polymer* **34**, 1571 (1993).
6. L. Athouël, G. Froyer and M.T. Riou, *Synth. Met.* **55-57**, 4734 (1993).
7. R. Resel, N. Koch, F. Meghdadi, G. Leising, W. Unzog and K. Reichmann, *Thin Solid Films* **305**, 232 (1997).
8. R. Resel and G. Leising (1997). *Surface Science* **409**, 302.
9. K. Erlacher, R. Resel, J. Keckes, F. Meghdadi, G.Leising, *J. Cryst. Growth* (1999) (in print).
10. A. Niko, F. Meghdadi, C. Ambrosch-Draxl, P. Vogl, G. Leising, *Synth.Met.* **76**, 177 (1996).
11. A. Piaggi, G. Lanzani, G. Bongiovanni, M.A. Loi, A. Mura, W. Graupner, F. Meghdadi and G. Leising, *Optical Materials* **9**, 489 (1998).
12. V.D. Hildebrandt, S. Doyle, H. Fuess, G. Pfaff and P. Reynders, *Thin Solid Films* **304**, 204 (1997).
13. M. Quaas and H. Wulff, *Fresenius J. Anal. Chem.* **361**, 617 (1998).
14. P. Scardi, F.C. Matacotta, V.I. Dediu and L. Correra, *J. Mater. Res.* **12**, 18 (1997).
15. F.L. Shan, Z.M. Gao and Y.M. Wang, *Thin Solid Films* **324**, 162 (1998).
16. F. Meghdadi, S. Tasch, B. Winkler, W. Fischer, F. Stelzer and G. Leising, *Synth. Met.* **85**, 1441 (1997).
17. G.K. Williamson and W.H. Hall, *Acta Metall.* **1**, 22 (1953).
18. R. Dehlez, Th.H. Keijser and E.J. Mittemeijer, *Fresenius Z. Anal. Chem.* **312**, 1 (1982).
19. D. Balzar and S. Popovic, *J. Appl. Cryst.* **29**, 16 (1996).

THE VALENCE ELECTRONIC STRUCTURE OF
N-DOPED P-SEXIPHENYL

U. Theissl[1], E.J.W. List[1], N. Koch[1,2], A. Vollmer[3], S. Schrader[2], J.-J. Pireaux[4], G. Leising[1]

[1] Institut f. Festkörperphysik, TU-Graz, A-8010 Graz, Austria
[2] LS Physik kondensierter Materie, Universität Potsdam, D-14469 Potsdam, Germany
[3] Institut f. Physikalische u. Theoretische Chemie, Freie Universität Berlin, D-14195 Berlin Germany
[4] Laboratoire Interdisciplinaire de Spectroscopie Electronique - Facultés Universitaires Notre-Dame de la Paix, B-5000 Namur, Belgium

ABSTRACT

Thin films of p-sexiphenyl (6P) were doped with increasing amounts of potassium *in situ*, and the change in the valence electronic structure of 6P upon the alkali metal deposition was followed with ultraviolet photoelectron spectroscopy. We observe the evolution of new intra-gap emissions, which are attributed to the formation of bipolarons, even for very low doping concentrations. The low binding energy intra-gap emission exhibits a pronounced asymmetric lineshape, in contrast to the findings when cesium is used as dopant. In order to investigate whether this lineshape is due to different emissive electronic species in the bulk and on the surface of the 6P film the take-off angle for the photoelectrons was varied. As no change in the lineshape is found when going from normal to near-grazing emission we can exclude that charged 6P molecules in the bulk and on the surface yield different valence electronic spectra. Therefore, the characteristic lineshape of the low binding energy emission is proposed to be related to the interaction of the doped organic molecule with the different counterions.

INTRODUCTION

The possibility of varying the conductivity of organic conjugated materials over many orders of magnitude by n- or p-type doping, and the resulting effect on their electronic structure have attracted much interest [1,2]. A major issue is how to describe the mechanisms of charge storage and charge transport in these systems. The surplus charges introduced by doping significantly modify the local geometric structure of the organic molecules due to a strong electron-phonon coupling. If an alkali metal is used for doping, as in our case, the resulting electronic species are discussed within the framework of negative polarons (radical anions) and negative bipolarons (dianions). By definition, a negative polaron results from filling the former lowest unoccupied molecular orbital (LUMO) of a conjugated chain with one electron, leading to a half filled orbital. The consequence is a destabilization of the former highest occupied molecular orbital (HOMO) and a stabilization of the former LUMO, yielding new electronic states in the normally forbidden energy gap of the pristine material. A finite density of valence states (DOVS) is then expected to be found at the Fermi-energy E_F [3-5]. In the case of a bipolaron, the former LUMO is completely filled with two electrons, and E_F is expected to be located well above the occupied electronic levels. Quantum-chemical calculations show that the two new intra-gap states of the bipolaron are located deeper in the gap than the polaronic states [6,7]. A finite density of states at E_F in photoemission experiments performed with doped samples is generally attributed to the presence of polaron states. Conversely, experiments of doping bithiophene with Cs have been reported [6], in which a polaron to bipolaron transition takes place upon increasing dopant concentration: but no DOVS is observed at the position of E_F in the polaron spectra. This could be attributed to an incomplete final state screening in the (small) organic molecules. Therefore, the interpretation of the experimental data is not always straightforward.

In our experiment we investigate the change in the valence electronic region of p-sexiphenyl (6P) upon doping with potassium. This model molecule for poly(para-phenylene) exhibits strong electroluminescence, and therefore is a promising candidate for the application in organic light emitting devices [8-10]. Additionally, 6P has been the object of a wide variety of investigations, and vast knowledge on the physical properties of this molecule already exists [9,11-15].

In a recent experiment we observed a finite DOVS at E_F when exposing 6P to small amounts of potassium, employing ultraviolet photoelectron spectroscopy (UPS) [7]. For higher K concentration, E_F appeared located well above the occupied electronic states. By comparing the experimental data with results of quantum-chemical calculations, we propose an interpretation that bipolarons are formed exclusively throughout all investigated doping-levels, regardless the observation of DOVS at E_F for low doping regimes (details can be found in [7]). In another set of UPS measurements thin films of 6P were exposed to a Cs flux. Again, new intra-gap states due to n-doping were observed and attributed to bipolarons, but the lineshape of the low binding energy emission exhibited marked differences to the new emission observed for K-doping. Whereas for Cs the shape of this emission is purely Gaussian, the peak for K is strongly asymmetric. One possible explanation for this variance could be that the adsorption properties of K differ for the 6P surface from the bulk, resulting in different electronic species for the surface and the bulk of the thin organic film. Quantum-chemical calculations have shown that the geometry of the molecule/dopant complex has a large influence on the energy position of the doping-induced electronic states [16].

In this contribution we present investigations on whether the origin of the asymmetry of the intra-gap peak for K-doping is that the doped 6P molecules on the surface have different electronic emissions than 6P molecules in the bulk. This was done by evaporating K onto thin films of 6P in a stepwise manner, and recording the photoemission spectra after each step for a number of different take-off angles. By varying the take-off angle from normal to near-grazing emission we vary the information-content of our UPS experiment from bulk- to surface-enhanced, respectively.

EXPERIMENT

The UPS experiments with K-doping were performed at the TGM2 beamline at BESSY I, Berlin, equipped with an ADES 400 spectrometer. Thin films of 6P (the powder purchased from Tokyo Chemical Industries Co. Ltd., Japan) were evaporated *in situ* from a resistively heated pinhole source onto a Si wafer, which previously was covered with Ti *ex situ*. The film thickness was monitored with a quartz microbalance. Throughout the whole experiments the samples were kept at room temperature. Potassium was evaporated from a SAES dispenser in a stepwise manner, the pressure in the preparation chamber remaining below 5×10^{-9} mbar. The photoemission spectra were recorded after each consecutive step with varying take-off angles, the photon energy was 26 eV. The sample transfer from the preparation to the analysis chamber was carried out while maintaining the UHV conditions, the pressure in the analysis chamber being below 1×10^{-10} mbar.

The Cs-doping experiments were performed at the FLIPPER II beamline at HASYLAB, Hamburg [17]. The same preparation parameters were applied as mentioned above, except that a stainless steel plate was used as a substrate. Cs was evaporated from a SAES dispenser. At this same experimental station also K-doping experiments were performed, which are reported in detail in [7].

RESULTS AND DISCUSSION

Fig.1. represents the starting point of our considerations. It shows a close-up of the valence electron region of 6P films which were doped with K and Cs, respectively. One can see the new low binding energy emissions in the formerly empty energy-gap of the pristine organic material due to doping. The topmost K and Cs spectra correspond to a 100% doping level, which means in our case that further exposure to the alkali metal flux did not lead to further changes for the whole valence band features. The two lower K and Cs curves correspond to approximately 30% doping level, revealing no significant change in the emissions compared to 100% doping. The binding energy scale is arbitrarily set to zero at the maximum of the intra-gap emission for K-doping. The alignment of the individual spectra with respect to each other was chosen in a way that the strongly localized π-state of 6P at higher binding energy (not shown) do line up, as this emission is expected to be least affected by doping [7]. Clearly the marked difference in both binding energy and lineshape of the

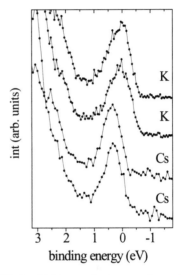

Fig.1. Valence electron region photo-emission spectra of K- and Cs-doped 6P films, recorded at HASYLAB

doping-induced new intra-gap emissions can be observed. In the case of doping with Cs the low binding energy peak is purely Gaussian, whereas the K-induced peak is strongly asymmetric with pronounced tailing towards higher binding energy. Also, the peak maximum for Cs is located at higher binding energy than for K. However, in both cases of doping experiments we concluded that the new electronic species formed were bipolarons [7,18]. These observations lead to the consideration that the adsorption properties (i.e. the geometry of the dopant-6P complex) of K are different for the organic film surface and bulk, resulting in different binding energies of the intra-gap peaks. This possibility is supported by Density Functional Theory calculations taking into account the interaction of a counterion with p-terphenyl (3P), where a strong dependence of the electron binding energies on the geometry of the counterion/3P complex were found [16].

In order to investigate whether the intra-gap emission of 6P due to K deposition is different for the film surface and bulk we varied the take-off angle (ToA) in a photoemission experiment, thereby changing the information-depth. Increasing amounts of K were deposited stepwise onto a nominally 200 Å thick 6P film. Fig.2. shows the results of this study for three different doping levels of (a) 5%, (b) 10%, and (c) 35%. The ToA is indicated in each figure;

0° correspond to normal emission. From these figures it becomes evident that the lineshape of the doping-induced gap-state does not change when going from normal to near grazing emission, i.e. from bulk- to surface-enhanced signal. The conclusion drawn from these observation is that in contrast to the expectations there is no difference in the photoemission from doped 6P molecules on the surface from those in the bulk, and that the asymmetry observed does not stem from different geometry-induced electronic states.

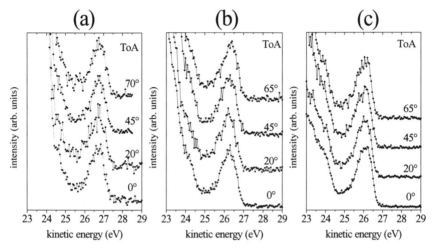

Fig.2. Valence electron region of K-doped 6P for various take-off angles (ToA) and doping levels: (a) 5%, (b) 10% and (c) 35%

Now it is interesting to compare the new intra-gap peaks at the same ToA for different doping levels. The reason is that one might expect that polarons (singly charged states) are formed at low doping levels, whereas the formation of bipolarons (doubly charged states) occurs at higher doping levels [3-5,6,19]. As the binding energy for the upper polaron and the upper bipolaron states are expected to be different, one would assume that the lineshape of the low binding energy intra-gap emission changes when going from the polaron regime to the bipolaron regime, which indeed has been observed in [6]. We mentioned earlier, that in another experiment where 6P was doped with K [7] we observed a finite density of valence states at E_F for doping levels below approximately 15%. For higher doping levels no DOVS were observed at E_F. From the movement of the Fermi-level alone one could conclude that polarons are present in the sample below 15% doping level and bipolarons above. Fig.3. shows the same spectra as Fig.2. but arranged in a different way. Each plot of Fig.3. corresponds to a certain ToA, namely (a) 0°, (b) 20°, (c) 45°, and (d) 65°. The curves in each plot were recorded at the different doping levels which were investigated: i) 5%, ii) 10%, and iii) 35%. No significant difference in the lineshape of the emission at the individual doping levels can be found. This is the case for all ToA's, therefore surface- and bulk-contributions to the signal are identical. The only conclusion that can be drawn from this comparison is that throughout all doping levels, regardless of the position of the Fermi-level found in [7], this intra-gap emission stems from the same electronic species. As for the completely doped 6P film [7,18] E_F is found above the DOVS the doping-induced emission is attributed to the presence of bipolarons. Therefore we must conclude in the present work that definitely no transition from polarons to bipolarons takes place when we increasingly dope 6P with K.

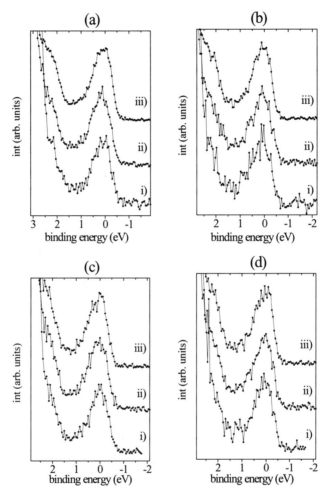

Fig.3. Valence electron region of K-doped 6P; the take-off angles are: (a) 0°, (b) 20°, (c) 45°, (d) 65°; the doping levels are; i) 5%, ii) 10%, iii) 35%; binding energies relative to the peak maximum

As we have shown that for a K-doped 6P film the photoemission signal from the surface and from the bulk are identical we postulate that the difference in the lineshape of the new intra-gap emission compared to doping with Cs stems from the interaction of the charged 6P molecule with the counterion. We rule out variances in the adsorption geometry of K on the 6P surface from that in the deeper lying layers. Therefore additional doping experiments with other dopants (like other alkali or alkaline earth metals) should help to explore the nature of the interaction between the conjugated organic material and the counterions.

CONCLUSIONS

Thin films of 6P were exposed stepwise to increasing amounts of K *in situ*. After each deposition step ultraviolet photoemission spectra were recorded with various take-off angles. New intra-gap states were observed due to n-doping of the organic molecules. In contrast to the findings for Cs-doping, where the intra-gap state was found to be one Gaussian peak, we observe a strongly asymmetric and broadened new peak in this work. The variation of the information depth has shown that the asymmetry of the doping-induced gap-states is not due to different electronic states on the surface and in the bulk of the 6P film. We conclude that the variance in the lineshape of the intra-gap emissions is due to a different interaction between the charged 6P molecules and the individual counterions.

ACKNOWLEDGEMENT

We acknowledge financial support of the EC-TMR project "*EUROLED*". EJWL's research is funded by the Austrian *FWF*, contract no. P-12806 PHY.

REFERENCES

[1] R. Menon, C.O. Yoon, D. Moses, A.J. Heeger, in Handbook of Conducting Polymers, 2nd ed., Eds T. Skotheim, R. Elsenbaumer, J. Reynolds, Dekker, New York 1997
[2] A.J. Heeger, S. Kivelson, J.R. Schrieffer, W.P. Su, Rev.Mod.Phys. **60**, 781 (1988)
[3] D. Baeriswyl, D.K. Campbell, S. Mazumdar, in Conjugated Conducting Polymers, Eds. H. Kiess, Springer Series in Solid-State Sciences **102** (1992)
[4] E.M. Conwell, H.A. Mizes, Phys. Rev. B **44**, 937 (1991)
[5] W.R. Salaneck, J.L. Brédas, Synth.Met. **67**,15 (1994)
[6] D. Steinmüller, M.G. Ramsey, F.P. Netzer, Phys. Rev. B **47**, 1323 (1993)
[7] N. Koch, L.-M. Yu, E. Zojer, R.L. Johnson, G. Leising, J.-J. Pireaux, J.-L. Brédas, J.Chem.Phys., submitted
[8] W. Graupner, G. Grem, F. Meghdadi, Ch. Paar, G. Leising, U. Scherf, K. Müllen, W. Fischer, F. Stelzer, Mol. Cryst. Liqu. Cryst. **256**, 549 (1994)
[9] G. Leising, S. Tasch, W. Graupner, in Handbook of Conducting Polymers, 2nd ed., Eds. T. Skotheim, R. Elsenbaumer, J. Reynolds, (Dekker, New York 1997)
[10] N. Koch, A. Pogantsch, E.J.W. List, R.I.R. Blyth, M.G. Ramsey, F.P. Netzer, G. Leising, Appl. Phys. Lett., in print
[11] R.Resel, N.Koch, F.Meghdadi, G.Leising, W.Unzog, K.Reichmann, Thin Solid Films **305** (1997) 232
[12] E.Zojer, J.Cornil, G.Leising, J.L. Brédas, Phys.Rev.B **59**, 1 (1999)
[13] S. Narioka, H. Ishii, K. Edamatsu, K. Kamiya, S. Hasegawa, T. Ohta, N. Ueno, K. Seki, Phys. Rev. B **52**, 2362 (1995)
[14] K. Seki, U.O. Karlsson, R. Engelhardt, E.E. Koch, W. Schmidt, Chem. Phys. **91**, 459 (1984)
[15] N. Koch, L.M. Yu, V. Parenté, R. Lazzaroni, R.L. Johnson, G. Leising, J.J. Pireaux, J.L. Brédas, Adv. Mater. **10**, 1038 (1998)
[16] E. Zojer, private communication
[17] R.L. Johnson, J. Reichardt, Nucl. Instr. Methods **208**, 719 (1983)
[18] N. Koch, L.M. Yu, A. Rajagopal, R. L. Johnson, G. Leising, J.J. Pireaux, J.Vac.Sci.Technol., submitted
[19] G. Iucci, K. Xing, M. Lögdlund, M. Fahlman, W.R. Salaneck, Chem. Phys. Lett. **244**, 139 (1995)

EL BEHAVIOR OF STYLYL COMPOUNDS WITH BENZOXAZOLE AND BENZOTHIAZOLE FOR ORGANIC LIGHT-EMITTING-DIODE

Koichi Yamashita*, Kaname Imaizumi*, Tatsuo Mori* and Teruyoshi Mizutani* and Hiroshi Miyazaki**
* Department of Electrical Engineering, Graduate school of Engineering, Nagoya University, Furo-cho, chikusa-ku, Nagoya, 464-8603 JAPAN
k-yamasi@echo.nuee.nagoya-u.ac.jp
**Nippon Steel Chemical Co., Ltd., Nakabaru, Tobata-ku, Kitakyushu, 804-8503 JAPAN

ABSTRACT

We studied the EL behaviors of benzoheterocyclic derivatives as emitting and hole transport layers for organic light-emitting-diodes (OLEDs). First, we studied the benzoheterocyclic derivatives having stylyl group, triphenylamine group and benzoxazole or benzothiazole group as an emission layer. These devices emitted a blue-green light. The current densities of the OLED having these benzoheterocyclic derivatives as an emission layer were higher than that of the Alq3 OLED at same applied voltage. However, these devices did not have a high EL efficiency (maximum 0.1 lm/W). From these results, in these benzoheterocyclic derivatives having a triphenylamine group, we thought that holes could flow out from hole transport to cathode. We newly synthesized dimer-type benzoheterocyclic derivatives without a triphenylamine group, which have benzoheterocycle at both sides. The current densities of the OLEDs having dimer-type benzoheterocyclic derivatives was more strongly suppressed than that of the OLEDs having benzoheterocyclic derivatives with a triphenylamine group at same applied voltage, but the EL efficiency could not be improved by dimerization and eliminating of a triphenyldiamine group obtained. Next, we studied the TPD derivatives having benzoxazole, benzothiazole and stylyl groups as hole transport layer. In new TPD derivatives, the EL efficiency the OLEDs having the TPD derivatives with stylyl groups was the best efficiency of all. The EL efficiency of ITO/a TPD derivative with stylyl groups/Alq3/AlLi is 1.1 lm/W (max. luminance $12000cd/m^2$).

INTRODUCTION

An organic light-emitting-diode (OLED) has been shown to have a high luminance, a high efficiency and a long lifetime as a practical monochromatic EL display [1]. There are a lot of studies an emission materials and carrier transport materials for OLED. The studies of organic metal complexes and non-metal organic compounds as an emission layer have been reported by C. W. Tang et al., J. Kido et al., S. Tokito et al. and Y. Hamada et al., [2-5]. The use of distylylbenzene derivatives for a practical blue-emitting OLED was first discovered by Hosokawa et al., [6].

In this paper, we studied on two series of new stylyl compounds as an emitting or a hole transport material for OLEDs. One is a benzoheterocyclic derivatives having a triphenylamine and a benzoheterocyclic group which is a benzoxazole or a benzothiazole group. Furthermore, benzoheterocyclic derivatives eliminated with triphenylamine groups are newly synthesized. Another has the structure of ditriphenylamine derivative (TPD) having benzoxazole, benzothiazole and stylyl groups. We estimated EL properties of the OLED consisting of these materials.

EXPERIMENTAL

Figure 1 shows our device structure. Our device has a bilayer organic structure (anode / hole transport layers (HTL) [50nm] / emission layer (EL) [50nm] / cathode. The anode and cathode

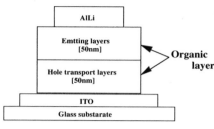

Figure 1. Our OLED structure.

are ITO and AlLi alloy(Li 0.05wt%), respectively. In the case of using our materials as an emission layer, we used N,N -diphenyl-N,N - bis(3-methylphenyl)-1,1 diphenyl-4,4 -diamine (TPD) as a hole transport layer. In the case of using our materials as a hole transport layer, we used 8-hydroxyquinoline aluminum (Alq3) as an emission material. Organic thin films and electrodes were formed by means of the vacuum vapor deposition method. The details of fabrication are shown in the previous paper [7]. The current-voltage (I-V) characteristics of organic LEDs were measured on a programmable electrometer having current and voltage sources Source Measure Unit, model 237, (Keithley). The EL spectra were measured by out on an optical multichannel analyzer PMA-10, (Hamamatsu Photonics). The luminance was measured with a luminance meter BM-8, (Topcon). The I-V, EL spectrum and luminance measurement were carried out in vacuum at room temperature.

RESULTS AND DISCUSSION

The OLEDs having benzoheterocyclic derivatives with a triphenylamine group as an emission layer

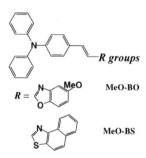

Figure 2. Chemical structures of benzoheterocyclic derivatives having a triphenylamine group.

Figure 2 shows the chemical structures of our benzoheterocyclic derivatives. These benzoheterocyclic (a benzoxazole and a benzothiazole) derivatives have triphenylamine group as side group. All OLEDs having these materials as an emission layer emitted a radiation from each material. For example, the luminance of OLED consisting from ITO/TPD[50nm]/Phe-Bo[50nm]/ AlLi was about 300 cd/cm^2 at 10 V. Figures 3 and 4 show PL spectra of benzoheterocyclic derivative thin films and EL spectra, respectively. The substituent on heterocycle did not influence an emission color from OLED. Figure 5 and 6 show the current density-voltage characteristics and luminance-current density characteristics, respectively. We showed the characteristics of the OLED having Alq3 as

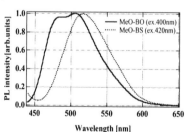

Figure 3. PL spectra of benzoheterocyclic derivative thin films.

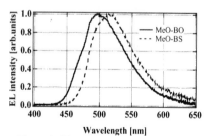

Figure 4. EL spectra of the OLEDs having benzoheterocyclic derivative emission layer.

an emission layer as a reference device. The current densities of all OLEDs having these benzoheterocyclic derivatives are higher than that of the Alq3 OLED. However, the luminance per current density for these OLEDs are one order lower than that of the Alq3 OLED. Figure 7 shows the energy diagram of benzoheterocyclic derivatives. The ionization potential (Ip) was estimated by AC-1, (Riken Keiki). The optical band gap was estimated from an optical absorption. The Ip of MeO-BO and MeO-BS are lower than that of Alq3. Therefore, the effect of hole blocking for these benzoheterocyclic derivatives are weaker than that for Alq3 in interface between TPD and an emission layer. From these results, we think that the OLEDs having these benzoheterocyclic derivatives as an emission layer have much current component without contributing recombination, which holes injected from ITO anode into emission layer through TPD layer flow out to AlLi cathode. We consider that these phenomena are caused by a triphenylamine group which act as hole transport site, that is, benzoheterocyclic derivatives have a good hole transportation.

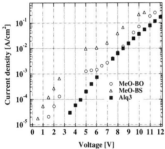

Figure 5. The current density-voltage characteristics of OLEDs having benzoheterocyclic derivatives.

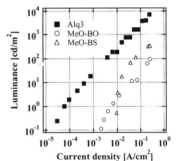

Figure 6. The luminance-current density characteristics of the OLEDs having benzoheterocyclic derivatives.

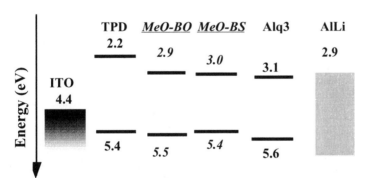

Figure 7. The energy diagram of benzoheterocyclic derivatives.

The EL characteristics of dimer-type benzoheterocyclic derivatives

From the above result, we designed the following chemical structures with dibenzoheterocyclic group and without triphenylamine group (Figure 8). Figure 9 shows our molecular design concept. These stylyl derivatives are dimer-type benzoheterocyclic derivatives having two heterocyclic ring groups eliminated triphenylamine group. The dimer-type benzoheterocyclic derivatives have a stronger PL intensity than Alq3 in thin films.

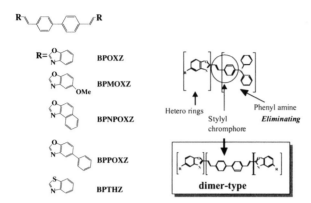

Figure 8. Chemical structures of the dimer-type stylyl derivatives.

Figure 9. Molecular designing concept.

The EL emission was observed in the BPOXZ and BPTHZ OLEDs. Figure 10 shows the current density-voltage characteristics of the BPOXZ and BPTHZ OLEDs. The current density of the OLED having BPOXZ or BPTHZ emission layer are lower than that of the Alq3 OLED. The current densities of the dimer-type benzoheterocyclic derivatives without a triphenylamine group was more strongly suppressed than those of monomer-type benzoheterocyclic derivatives with a triphenylamine group. Figure 11 shows the luminance-current density characteristics of the BPOXZ and BPTHZ OLEDs. The luminance per current density of the OLEDs with dimer-type

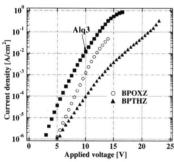

Figure 10. The current density-voltage characteristics of the BPOXZ and BPTHZ OLEDs.

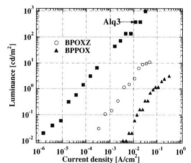

Figure 11. The luminance-current density characteristics of EL emitted devices.

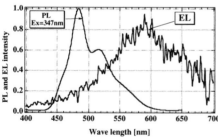

Figure 12. The normalized EL spectrum of OLED and normalized PL spectra of BPTHZ thin layers.

of benzoheterocyclic derivatives were lower than that of the OLEDs with Alq3 and with benzoheterocyclic derivatives having a triphenylamine group. Consequently, the dimerization and elimination of a triphenylamine group could suppress current density including flowing hole current, but they could not improve the EL efficiency.

Figure 12 shows the EL spectrum of the BPTHZ OLED and the PL spectra of BPTHZ thin film. The EL spectrum of the BPTHZ OLED did not agree with the PL spectrum of BPTHZ thin film. The EL spectrum of the BPTHZ shifted to the long wavelength region. The apparent spectrum-shift may be explained by that a high-energetic EL component is diminished whether by the formation of the exciplex between TPD and BPTHZ molecules or by concentration quenching.

The EL characteristics of the OLED having new TPD derivatives as a hole transport layer

Next, we studied new TPD derivatives as a hole transport layer for OLEDs. Figure 13 illustrates the chemical structures of TPD and its derivatives. These derivatives have distylyl (STPDPH), benzoxazole (STPBO), methoxybenzoxazole (STPMEOBO) and naphtothiazole (STPNAPTH) groups on each aromatic ring of triphenylamine group, respectively. The glass transition temperature (Tg) of TPD is estimated to be 64 °C. It is well known that TPD is an excellent hole transport material but the polycrstalization rate of TPD thin film is very fast. The Tg of new STPDPH is estimated to be 90 °C. This thin film could not polycrystalize for several month at room temperature.

Figure 14 shows the luminance-voltage characteristics of the OLEDs having these TPD derivatives as a hole transport layer. The OLEDs consist of ITO/TPD derivatives [50nm]/Alq3[50nm]/AlLi. The luminance of the OLEDs with STPBO, STPMEOBO and STPNAPTH are lower than that with TPD at a same applied voltage, whereas that of the OLED with STPDPH is the same as latter. Table 1 summarizes Tg, maximum power efficiency and maximum luminance efficiency of OLEDs having these TPD derivatives as a hole transport layer. In these TPD derivatives, the EL efficiency of the OLED with STPDPH is the best of all. The EL efficiency of the OLED with STPDPH (1.1 lm/W) is lower than that of the OLED with TPD (1.3 lm/W). The ionization potential (Ip) of TPD and STPDPH estimated by UPS method are 5.33 eV and 5.34 eV, respectively. Therefore, there is no difference between both barrier heights of hole injection from ITO anode.

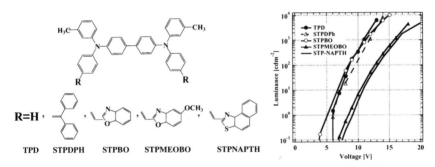

Figure 13. Chemical structures of TPD derivatives as hole transport materials.

Figure 14. The luminance-voltage characteristics of the OLED consisting of TPD derivatives and Alq3.

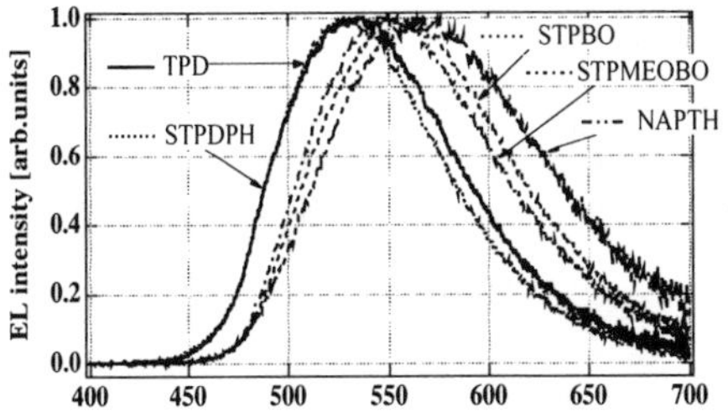

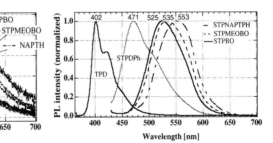

Figure 15. EL spectra of the OLEDs with TPD derivatives as a hole transport layers.

Figure 16. PL spectra of thin films with TPD derivatives.

Table 1. Tg, maximum power efficiency and maximum luminance efficiency of OLEDs having these TPD derivatives.

Specimen	Tg (°C)	Maximum (lm/W)	Maximum cd/A
TPD	63.4	1.3	3.0
STPDPH	90.0	0.8	2.6
STPBO	No measure	0.5	1.4
STPMEOBO	No measure	0.1	0.5
STPNAPH	No measure	0.05	0.3

Figure 15 shows the EL spectra of the OLEDs consisting of the TPD derivatives and Alq3. The EL spectra of the OLED with STPDPH as a hole transport layer agreed with that of the OLED with TPD layer. These devices emitted a radiation from Alq3. However, the EL spectra of the others devices with TPD derivatives shifted to long wavelength region. Figure 16 shows PL spectra of TPD derivatives. In the case of using the TPD derivatives except for STPDPH as a hole transport layer, the emitting zone is not in only Alq3 layer and is extended from the TPD derivative layer to Alq3 layer. Consequently, both EL from Alq3 and the TPD derivatives can be observed in EL spectra.

CONCLUSIONS

We studied the characteristics of benzoheterocyclic derivatives as an emission or a hole transport layer for OLEDs. The OLEDs having benzoheterocyclic derivatives with a triphenylamine as an emission layer have about maximum luminance $300cd/m^2$ at 10 V. However, the luminance of these OLEDs were 1-2 orders lower than that of the Alq3 OLED at a same applied voltage. We think that the OLEDs having these benzoheterocyclic derivatives as an emission layer have much current component without contributing recombination. Hole carrier injected from ITO anode into an emission layer (benzoheterocyclic derivatives having triphenylamine group) through TPD layer flow out to AlLi cathode. The OLEDs having dimer-type benzoheterocyclic derivatives without triphenylamine group eliminated were suppressed current density. The maximum luminance of these OLEDs having dimer-type heterocyclic derivative were about $20cd/m^2$. The EL spectra of benzoheterocyclic derivatives did not agree with PL spectra of benzoheterocyclic derivative. The OLED with STPDPH as a hole transport layer can have the same luminance per current density as the OLED with TPD. It was found that STPDPH thin film has stabler film structure than TPD thin film.

REFERENCES

1. T. Wakimoto, R. Murayama, K. Nagamiya, Y. Okuda, H. Nakada and T. Tohma, Society for Information Display 96 Digest, 849 (1996).
2. C. W. Tang and S. A. Vanslyke, Appl. Phys. Lett., **51**, 913 (1987).
3. J. Kido and J. Endo, Chem. Lett., 593 (1997).
4. S. Tokitou, H. Tanaka, Y. Taga, and A. Okada, J. Mater. Chem., **8**, 1999 (1998).
5. Y. Hamada, T. Sano, H. Fujii, Y. Nishio, Jpn, J. Appl, Phys., **35**, 1341 (1996).
6. C. Hosokawa, M. Eida, M. Matsuura, K. Fukuoka, H. Nakamura and T. Kusumoto; Synth. Met., **91**, 3 (1997).
7. T. Mori, K. Miyachi and T. Mizutani, J. Phys. D, **28**, 1461 (1995).

IMPROVING THE CRYSTALLIZATION RESISTANCE OF ALUMINUM(III) 8-HYDROXYQUINOLINE-BASED EMITTING MATERIALS BY ENTROPIC STABILIZATION

KEITH A. HIGGINSON, BAOCHENG YANG, AND FOTIOS PAPADIMITRAKOPOULOS*
Nanomaterials and Optoelectronics Laboratory, Department of Chemistry, Polymer Science Program, Institute of Materials Science, University of Connecticut, Storrs, CT 06269-3136

ABSTRACT

The morphological stability of evaporated films of aluminum(III) 8-hydroxyquinoline (Alq_3) was investigated. Films which were found to be non-crystalline by x-ray diffraction upon deposition, crystallized rapidly upon annealing, especially where defects were present. Blends of Alq_3 with aluminum(III) 5-methyl-8-hydroxyquinoline were proposed for thermally stable amorphous emitting layers in light-emitting diodes. Films coevaporated at a 1:1 ratio did not show evidence of crystallization or phase separation even after long annealing periods at temperatures as high as 160 °C.

INTRODUCTION

Since 1987, when electroluminescence of aluminum(III) 8-hydroxyquinoline (Alq_3) films was found to occur at low bias,[1] small-molecule and polymeric organic materials have been pursued as low-cost and versatile alternatives to existing electroluminescence technology.[1-5] At present, a number of companies have demonstrated small monochrome displays based on this technology, with Pioneer taking the lead in commercializing an 64×256 passively addressed monochrome electroluminescent display based on Alq_3 technology for use in car stereos.[6] A real barrier to marketing organic light-emitting diode (OLED) technology, however, has been achieving devices capable of sustaining necessary brightness levels with minimum drift over long periods of time, as well as maintaining longevity while operating at elevated temperatures (e.g., 60-80 °C). Typically, devices show failure locally, evidenced by origination and growth of nonemissive regions. In parallel, a gradual decrease in brightness, accompanied by an increase in driving voltage, also occurs. Several failure modes describing this behavior have been enumerated, and may be categorized generally in two groups: chemical failures, where either the cathode or organic material suffers degradation due to sorption of water or oxygen, and/or formation of charged or excited species; or physical change, due to poor adhesion between metal and organic, diffusion of organics and anode elements, and crystallization of organic materials.[3]

It is the crystallization of the component films which is the subject of this paper. Current design strategies for thermally stable devices employ hole transport materials with high glass transition temperatures (T_g),[7-9] principally by increasing the molecular weight and asymmetry of the components. This approach reaches practical limits from both a purification and fabrication standpoint, as recrystallization and sublimation become problematic as the size of the molecules is increased. While an adequate rule of thumb, it may not always be the ideal design strategy for thermally stable components. Alq_3, as a pointed example, has a very high T_g, but will crystallize very rapidly at significantly lower temperatures, a property which it does not share with readily glass-forming hole transporting materials such as TPD. The rate of crystallization (or devitrification) of amorphous Alq_3 is negligible near room termperature, but may be significant over long times, in locally heated regions, or in large area devices if temperature increases are as high as have been postulated.[10]

Glass forming ability is a term which describes the critical cooling rate necessary to form a stable glass from the liquid state, or alternatively, the critical isothermal annealing time needed to devitrify an existing glass.[11-13] The latter definition represents the stability of an amorphous film against crystallization, and is of principle interest when designing materials for OLEDs. Increasing the glass forming ability is synonymous with retarding the crystallization rate. The rates for both nucleation and growth are governed by both a thermodynamic driving force for crystallization in addition to an energy barrier for molecular rearrangements, although the latter is solely considered for high-T_g design. These processes tend to be competitive for crystallization: at low temperatures (well below T_g) the viscosity becomes large and the rate is controlled by molecular motion; at higher temperatures (above T_g and approaching the equilibrium melting point, T_m, where the system resembles an undercooled liquid) the driving force for crystallization is small and tends to be rate-limiting. Selecting materials with high activation energies for rearrangement, or which vitrify at low undercoolings (and have a correspondingly low driving force), will act to reduce the total rate of crystallization of a film.

The experimentally observed glass transition temperature is a measure of the resistance to molecular motion, indicating components which relax at the same time scale of the experiment. Materials with high T_gs have a large barrier for cooperative rearrangements. According to the theory of Gibbs and DiMarzio, between the experimentally observed T_g and the true second-order transition temperature T_2, there is a finite probability for cooperative rearrangements of molecules or segments to occur. A rule of thumb developed by Adam and Gibbs for polymeric or low molecular weight systems is $T_g/T_2 \approx 1.3$.[11] By this, Alq$_3$ ($T_g = 175$ °C) is perfectly frozen below about 75 °C and forms a stable glass at room temperature. For TPD, a commonly used hole transporter ($T_g = 65$ °C), T_2 is at about -13 °C and this material eventually crystallizes at ambient temperatures.

On the other hand, the thermodynamic parameter affecting nucleation and growth rates is the Gibbs free energy difference between the crystalline and liquid or glassy phase (ΔG_{gc}). Increasing T_g is a valid approach for the design of thermally stable materials, but this is not necessarily sufficient to achieve high glass forming ability. Another strategy is to minimize the value of ΔG_{gc}. Naito, based on this type of analysis, suggests designing organic dyes with minimal changes in enthalpy for glass-crystal transitions (ΔH_{gc}) and low equilibrium melting points (T_m), in addition to high T_gs.[14,15] With similar reasoning, a small difference between T_g and T_m is often considered a useful indicator for high glass forming ability.[13,16]

The introduction of miscible impurities can be used to increase the glass forming ability of liquid materials. For non-polymeric or non-network liquids such as metals, the necessary impurity levels may be as low as 1-2%.[13,17] Mixing makes the liquid state more thermodynamically favorable by increasing its entropy, lowering the energy of the solution with respect to either single component. For an ideal solution, the free energy change of mixing multiple components is equal to the combinatorial entropy obtained, although deviations from ideality (due to intercomponent enthalpic interactions) generally occur. The commonly observed effect of mixing is the depression of the equilibrium melting point. Since the free energy of a miscible blend is lower than either that of either component, ΔG_{gc} is smaller than in the single component glass at the same temperature, and the driving force for crystallization is reduced. Furthermore, slight negative deviations from ideality (due to attractive intermolecular forces) can lower the free energy of the mixture even more. For simple binary systems, stable glass forming regions are usually centered about deep eutectics, where the free energy of mixing is largest.[13]

It should be noted that these arguments is fairly qualitative. It is significantly more difficult to mathematically describe crystallization kinetics when the crystal and liquid phase have different compositions, as is normally the case for multicomponent systems.[12] Model parameters for mixture properties are fairly accessible, although at present we are designing blends of electronically similar compounds in an effort to preserve charge conduction and emissive properties. Under this umbrella, we have formulated blends of chemically similar emitting materials (approximating an ideal solution) at a 1:1 mole ratio (maximizing the combinatorial entropy). Aluminum(III) 4-methyl-8-hydroxyquinoline (AlMq$_3$) is a blue-green emitting material which has received recent attention for OLED applications.[18] Blends of AlMq$_3$ and Alq$_3$ were made by depositing the two materials from the same evaporation source. The deposition rate was fairly fast (about 20 Å/s) in order to preserve the 1:1 ratio. Current efforts are underway to characterize the T_g (which should range between that of the two components; the T_g of vapor-deposited AlMq$_3$ is similar to that of Alq$_3$) and other important parameters of the blend.

EXPERIMENTAL

All films were prepared by physical vapor deposition in an Edwards Auto 306 vacuum coater equipped with a turbomolecular pump. Blends of Alq$_3$ and AlMq$_3$ were deposited from a single source at about 20 Å/s to approximately preserve the stoichiometry, while single-component films were evaporated at typical rates of about 5 Å/s. Alq$_3$ and AlMq$_3$ were synthesized by standard methods.

Wide angle X-ray diffraction (WAXD) was performed on evaporated material which was scraped from its substrate and sealed in a quartz capillary. Annealing of these samples was performed by placing the capillary in a constant temperature bath. WAXD patterns were obtained with a Bruker X-ray diffractometer with a GADDS area detector.

Polarized light and transmission optical microscopy was done performed on a Nikon Labophot transmission microscope.

Differential scanning calorimetric (DSC) measurements were conducted with a Perkin Elmer DSC-7, employing a 20 mL/min flow of dry nitrogen as a purge gas for the sample and reference cells. The temperature and power ordinates of the DSC were calibrated with respect to the known melting point and heat of fusion of a high purity indium standard. The scan rate was 10 °C/min. A two stage intercooler based on freon 502 and ethane 170 was used to quench the samples to -65 °C.

RESULTS AND DISCUSSION

As noted, Alq$_3$ is an example of a material with a high glass transition temperature, but of low glass forming ability.[14] In fact, little concensus has been reached in the literature as to the morphology of the evaporated films, ranging from purely amorphous[19] to semicrystalline.[1,20] Substrate temperature, deposition rate, and purity levels probably account for some of the discrepancies in the reported values. Figure 1a shows WAXD profiles of Alq$_3$. For a freshly sublimed sample, only a broad hump is visible, characteristic of a material with no long range order. With annealing, the trace rapidly resolves into crystalline peaks, even at temperatures 35 °C below its T_g. Crystallization of Alq$_3$ films was also observed with optical microscopy. 1000 Å films were annealed in a vacuum oven at the same temperature as the diffraction experiments. Crystallites at film defects form rapidly and grow with spherical symmetry. After an apparent induction time, smaller cylindrical or needle-like crystallites become observable in defect-free areas.

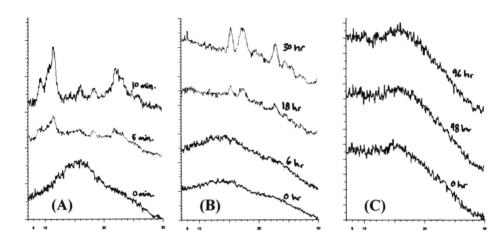

Figure 1. Wide angle X-ray diffraction patterns of (A) Alq₃ annealed at 145 °C, and (B) AlMq₃ and (C) 1:1 coevaporated blends of Alq₃ and AlMq₃ annealed at 160 °C.

Figure 2 shows differential scanning calorimetry (DSC) traces for freshly deposited Alq_3 films. Large exotherms above T_g (165 °C) are visible in the first heating scan of the newly sublimed sample. Below T_g erratic exotherms are also visible on the initial scan, which are ascribed to crystallization based on WAXD and microscopy data. Exposure to moisture, however, introduced impurities and eliminated these exotherms. A broad, featureless T_g at the reported value of about 175 °C is visible on repeated scans. The rapid low temperature crystallization is indicative of the abnormally high amount of free energy "frozen in" to the sublimed film. Annealing the sample in the vicinity of T_g allows the rearrangement of molecules into lower energy crystalline states. The remaining amorphous material is in a state of metastable equilibrium with the crystal phase (very much like a subcooled liquid), exhibiting a broad spectrum of relaxations associated with T_g.

The X-ray diffraction patterns of $AlMq_3$ and a 50% $Alq_3/AlMq_3$ blend are shown in Figure 1b and 1c as a function of annealing time. $AlMq_3$ demonstrated greater thermal stability than Alq_3, but still eventually resolved into crystalline peaks with sub-T_g annealing. The blend, however, did not crystallize even after one week of heating at 160 °C. Polarized light micrographs of the annealed films showed no birefringence (indicative of crystallite formation) as was the case for Alq_3. For X-ray diffraction, the film was immediately scraped and sealed in a capillary, but for the microscopy experiment, the film had a much greater exposure to atmospheric water and to residual organic materials in the vacuum oven where the annealing took place. The absorbed impurities may have acted to inhibit crystallization, as in sublimed Alq_3 where absorbed water eliminated the crystallization exotherms in the DSC.

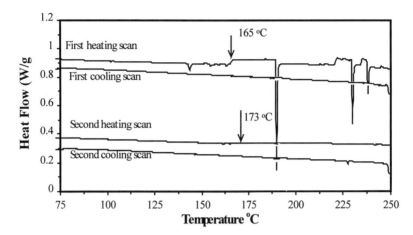

Figure 2. DSC trace for freshly sublimed Alq₃. Initial scan shows crystallization exotherms above and below T_g. Subsequent scans show no crystallization and a less distinct T_g.

Molecular doping has been often used to increase the quantum efficiency in Alq₃-based LEDs.[4] The observed lifetime enhancement also associated with this, may well be due to the suppression of crystallization by impurities, especially when noting the observations above. Addition of dopants to hole transport materials has already seen some success in retarding crystallization.[7] Adding a higher percentage of electronically similar material to the emitting layer (with or without the addition of other dopants) is expected to enhance the operational stability of OLEDs more effectively.

CONCLUSIONS

The morphological stability of Alq₃ is called into question, possibly as a cause of gradual "intrinsic" performance decay of OLEDs, especially when such devices are exposed to high temperatures or where heat dissipation is poor. If the temperature range for crystallization is approached (*i.e.*, near 80 °C), the crystallization of Alq₃ will be much more rapid than other materials in a corresponding state. Alq₃ was "entropically stabilized" by blending with AlMq₃ at 50 mol%, where maximum free energy change on mixing—and hence the slowest crystallization rate—should occur as a function of composition at first approximation. These blends were found to exhibit superb stability against the pure parent compounds by several orders of magnitude.

REFERENCES

1. C. W. Tang and S. A. VanSlyke, App. Phys. Lett. **51,** 913 (1987).

2. L. J. Rothberg and A. J. Lovinger, J. Mat. Res. **11,** 3174-3187 (1996).

3. J. R. Sheats, H. Antoniadis, M. Hueschen, W. Leonard, J. Miller, R. Moon, D. Roitman, and A. Stocking, Science **273,** 884 (1996).

4. P. E. Burrows, S. R. Forrest, and M. E. Thompson, Current Opinion Solid State Materials Science **2**, 236 (1997).

5. D. Bradley, Current Opinion Solid State Materials Science **1**, 789 (1996).

6. T. Tohma, in *International Display Research Conference Proceedings*, 1997, p. F-1.

7. B. E. Koene, D. E. Loy, and M. E. Thompson, Chem. Mat. **10**, 2235-2250 (1998).

8. Y. Shirota, Y. Kuwabara, H. Inada, T. Wakimoto, H. Nakada, Y. Yonemoto, S. Kawami, and K. Imai, App. Phys. Lett. **65**, 807 (1994).

9. S. Tokito, H. Tanaka, K. Noda, A. Okada, and Y. Taga, IEEE Trans. Elec. Dev. **44**, 1239-1244 (1997).

10. J. C. Sturm, W. Wilson, and M. Iodice, IEEE Journal of Selected Topics in Quantum Electronics **4**, 75-82 (1998).

11. J. Zarzycki, *Glasses and the Vitreous State* (Cambridge Universtiy Press, Cambridge, UK, 1991).

12. M. G. Scott, in *Amorphous Metallic Alloys*, edited by F. E. Luborsky (Butterworths, London, 1983), p. 144-168.

13. H. A. Davies, in *Amorphous Metallic Alloys*, edited by F. E. Luborsky (Butterworths, London, 1983), p. 8-25.

14. K. Naito and A. Miura, J. Phys. Chem. **97**, 6240-6248 (1993).

15. K. Naito, Chem. Mater. **6**, 2343-2350 (1994).

16. W. J. MacKnight, F. E. Karasz, and J. R. Fried, in *Polymer Blends*; Vol. *1*, edited by D. R. Paul and S. Newman (Academic Press, San Diego, 1978), p. 186-243.

17. P. F. James, in *Nucleation and Crystallization in Glasses*; Vol. *4*, edited by J. H. Simmons, D. R. Uhlmann, and G. H. Beall (American Chemical Society, Columbus, OH, 1982), p. 1-47.

18. J. Kido, Chem. Lett., 963-964 (1997).

19. T. Wada, Y. Yogo, I. Kikuma, M. Masui, M. Anzai, and M. Takeuchi, App. Surface Sci. **65/66**, 376-380 (1993).

20. K. A. Higginson, X.-M. Zhang, and F. Papadimitrakopoulos, Chem. Mater. **10**, 1017-1020 (1997).

CURRENT-VOLTAGE CHARACTERISTICS OF SINGLE LAYER POLYMER LIGHT EMITTING DIODES UNDER HIGH INTENSITY ELECTRICAL EXCITATION

B. RUHSTALLER*, J.C. SCOTT**, P.J. BROCK**, S.A. CARTER*
* Dept. of Physics, University of California, Santa Cruz, CA 95064
** IBM Almaden Research Center, San Jose, CA

ABSTRACT

We study the current-voltage characteristics of both double and single carrier polymer light emitting devices in the high intensity electrical excitation regime. Single layer devices are investigated with the orange-emitting MEH-PPV as the active polymer. Performing very low duty (≈ 0.001 %) pulsed electroluminescence measurements with electric fields up to $1*10^9$ V/m, we observe for the hole-dominated devices a space charge limited current with a saturating mobility and a saturating external quantum efficiency of 1%. For double carrier devices, the current starts to saturate in the high-field regime, appoaching Ohmic-like behavior. We report current densities of 100 A/cm^2 and brightnesses of $7*10^5$ cd/m^2. The spectral features are monitored below the onset of degradation.

INTRODUCTION

The high field regime in electrically pumped polymer LEDs is still poorly understood. It has first been studied by Braun et al [1] and just recently has regained attention by Tessler et al [2, 3]. To explore this regime very low duty pulses are required in order to avoid current heating [1, 3] and degradation [4].

In the intermediate field regime ($\leq 10^8$) using Ohmic contacts, high-purity MEH-PPV is known to have trap-free charge transport which can be described by the single carrier space charge limited (SCL) current formula with a field dependent mobility [5]. Both the field dependence coefficient γ and the zero field mobility μ_0 can be determined for electron-dominated as well as for hole-dominated devices [5]. The blue-emitting ladder type polymer Me-LPPP shows qualitatively similar field dependence in the intermediate field regime [6].

In this paper we present experimental data in the current regime up to 100 A/cm^2 and model the current with the unipolar SCL current formula.

EXPERIMENT

We use ITO patterned glass substrates on which a layer of a conducting polymer (PAni) is spun and dried in a vacuum oven. The polymers are spun cast onto these substrates from 1% wt solutions in p-Xylene to form 100nm thick films, and Ca and Al cathodes are thermally evaporated on top. Three different instrumentations are employed to cover the wide voltage regime. In the low field regime a Keithley source-measure unit is used to acquire current versus voltage curves. The intermediate field regime is measured using a programmable HP Universal Source 3245 with 5 % duty and 5 kHz pulses where the current is measured using a digital SR 830 Lock In amplifier. The high field regime is explored with a Velonex 350 high power pulser at very low duty cycles (≈ 0.001 %), low repetition rates ($\approx$ 30 Hz) and pulse lengths on the order

185

of a few hundred nanoseconds. In this regime the current is read from the scope display of the voltage across a series resistance. The acquired data is plotted as a continuous curve covering the three regimes (as in Fig. 1).

RESULTS

The space charge limited current formula for unipolar transport $J(E) = 9/8 \; \varepsilon\varepsilon_0 \; \mu(E) \; E^2 \, /L$, where ε is the dielectric constant and L is the thickness, describes the intermediate field regime successfully using a field dependent mobility $\mu(E) = \mu_0 * \exp(\gamma * E^{1/2})$ [5]. As can be seen in the log-log plot of Fig. 1, in the high field regime the current is simply proportional to the square of the applied bias. This is equivalent to a simplified SCL formula with constant mobility. We extract a saturated mobility of $4*10^{-5}$ cm^2/Vs. The use of the unipolar SCL formula is justified by the low quantum efficiency of the hole dominated Al-cathode device (Fig. 1).

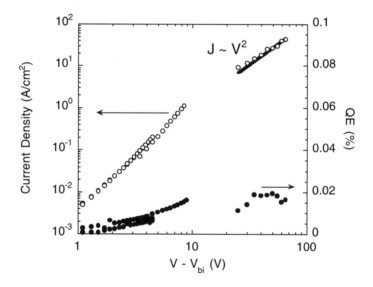

Fig. 1: Log-log plot of current (open symbols) and external quantum efficiency (closed symbols) versus voltage of a device with Al cathodes in the high field regime up to $8*10^8$ V/m. The voltage scale is corrected for a built in potential of 1.5 eV.

In Fig. 2, a comparison is shown of the luminance and quantum efficiency of devices with Ca and Al cathodes, respectively. We note that the quantum efficiencies do not drop significantly in the high field regime. The devices with Ca cathodes show higher currents and higher quantum efficiencies (up to 1.5%). In this case we find saturating current behavior in the high field regime. We achieve peak luminance of $7*10^5$ cd/m^2 and $4*10^5$ cd/m^2 for Ca and Al cathodes, respectively, at a bias of 70V. For Me-LPPP, we achieve a similar peak luminance of $5*10^5$ cd/m^2 at a bias of 40V using Ca cathodes..

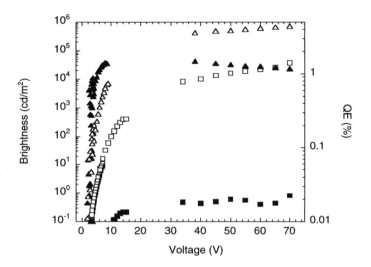

Fig. 2 Luminance (open symbols) and quantum efficiency (QE, closed symbols) versus voltage for devices with Ca (triangles) and Al (squares) cathodes, respectively.

No spectral changes are observed for biases up to 70V as shown in Fig. 3 for a Ca cathode device. Therefore heating effects [1, 3] can be excluded in our study at very low duty cycle.

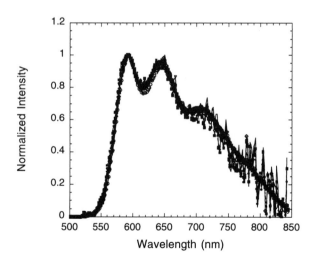

Fig. 3 Normalized spectra for different biases from 4 up to 70 Volts. No spectral changes are observed.

CONCLUSIONS

Both for the hole dominated as well as the double carrier devices, we find that in the high-field regime the current no longer follows the space charge limited formula with a field-dependent mobility as found in the low-field regime [3]. Instead, for the hole-dominated devices, we find a space charge limited current with a saturated mobility of $4*10^{-5}$ cm^2/Vs. Due to its moderate quantum efficiency we argue that the SCL formula for unipolar transport is still applicable. In the case of the double carrier devices the mobility also saturates, and the current approaches Ohmic-like behavior. No analytical model is currently available for modeling double carrier systems with different mobilities in the presence of carrier recombination.

ACKNOWLEDGMENTS

We would like to thank U. Scherf from Max-Planck Inst. für Polymerforschung Mainz for supplying the Me-LPPP powder and L. Bozano for stimulating discussions. The authors acknowledge financial support from NSF GOALI Grant #DMR 9704177. B.R. is thankful for support by the Sunburst Fonds ETHZ.

REFERENCES

1. D. Braun, D. Moses, C. Zhang, A.J. Heeger, Appl. Phys. Lett. **61**, 3092, (1992)

2. N. Tessler, N.T. Harrison, R.H. Friend, Adv. Mat. **10**, 64, (1998)

3. N. Tessler, N.T. Harrison, D.S. Thomas, R.H. Friend, Appl. Phys. Lett. **73**, 732, (1998)

4. V. Bliznyuk, S.A. Carter, J.C. Scott, G. Klärner, R.D. Miller, D.C. Miller, Macromolecules **32**,361, (1999)

5. L. Bozano, S.A. Carter, J.C. Scott, G.G. Malliaras, Appl. Phys. Lett. **74**, 1132, (1999) and references therein.

6. B. Ruhstaller, L. Bozano, S.A. Carter (unpublished).

CONJUGATED POLY-P-PHENYLENE (PPP) FROM POLY(1,3-CYCLOHEXADIENE) (PCHD) HOMO- AND BLOCK COPOLYMERS: CONTROLLED PROCESSABILITY AND PROPERTIES

J. MAYS, K. HONG, Y.WANG, R.C. ADVINCULA
Department of Chemistry, University of Alabama at Birmingham,
Chemistry Building, Birmingham, AL 35294-1240

ABSTRACT

Conjugated polymers have been used as solid-state materials for a variety of opto-electronic applications. Thus, their processability in a number of device fabrication protocols is an important consideration. In this report, we indicate our results on the synthesis of conjugated poly-p-phenylene (PPP) materials derived from poly(1,3-cyclohexadiene) (PCHD). The precursor PCHD polymers were synthesized by living anionic polymerization to produce homo- and block copolymer configurations with polystyrene. We have used a variety of initiators, solvent, and temperature conditions to determine the right parameters for obtaining narrow MWD polymers. The conditions for polymerization determined the ratio of 1,2 and 1,4 isomers in the microstructure. We then proceeded with conversion to PPP derivatives using a dehydrogenation reaction with chloranil. Our results indicate a systematic conversion to a conjugated polymer with increased solubility and photoluminescence properties. The microstructure, MW and block copolymer composition affecting the processability and conversion properties dramatically.

Introduction

Poly(p-phenylene) (PPP) and its derivatives are an interesting class of conjugated polymers, which exhibit conductivity upon doping.[1] They also exhibit photo- or electro-luminescence upon excitation with light or application of a bias voltage in a diode configuration.[2] Pure PPP is an insoluble and infusible material, stable at high temperatures. Its thermal stability is an attractive property, considering the requirements for future opto-electronic applications. However, this thermal stability is accompanied by a lack of processability, which increases with longer PPP sequences. Leising and co-workers first reported blue emission for a PLED fabricated using PPP.[3] This involved the preparation of PPP using a precursor polymer route derived from free radical polymerization of an esterified cyclohexadienediol monomer. Other attempts include the synthesis of poly (1,3-cyclohexadiene) (PCHD) derivatives using Zeigler type catalysts[4] and organolithium compounds,[5] which can be converted to PPP by subsequent aromatization normally by heating to high temperatures or dehydrogenation reactions.

In this report, we indicate our initial results on the synthesis of PCHD by living anionic polymerization, block copolymer configuration, and subsequent conversion to PPP derivatives. We have used a variety of initiators, solvent, and temperature conditions to determine the right parameters for obtaining narrow MWD polymers. The conditions for polymerization determined the ratio of 1,2 and 1,4 isomers in the microstructure. We then proceeded with conversion to PPP derivatives using a dehydrogenation reaction with chloranil. Our results indicate a systematic conversion to a conjugated polymer with increased solubility and photoluminescence properties. The microstructure, MW and block copolymer composition affected the processability and conversion properties dramatically. Recently, interesting core-shell cylindrical morphology has been observed for the PCHD precursor polymers by TEM and SAXS techniques.[6]

EXPERIMENT

Materials. All reagents and chemicals were purchased from Aldrich Chemical Co. and were purified accordingly. 1,3-cyclohexadiene was purified by exposure to CaH_2, Na dispersion and n-BuLi (30 min 0°C) on the vacuum line. Methanol was degassed on a vacuum line. The benzene and hexane solvents were treated with concentrated sulfuric acid for at least two weeks to remove olefinic impurities.

A variety of conditions were used to synthesize different derivatives. This involved the use of Butyllithiums (n-BuLi or s-BuLi) as initiators, followed by the addition of TMEDA or t-BuOK. Typical conditions for polymerization can be categorized as follows:
1. s-BuLi/t-BuOK or BenzylK: THF at -78 C/~24 hours
2. n-BuLi/TMEDA: cyclohexane at 25-50 C/ ~24 hours
3. s-BuLi: Methylcyclohexane at -78 C / ~168 hours

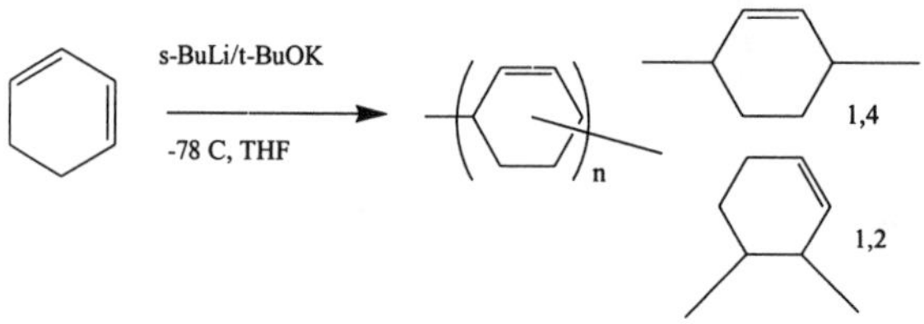

The reaction was carried out in a glass reaction apparatus (custom-made) consisting of the main reaction bulb and appropriate glass ampoules with break seals. Complete details of the synthesis procedure will be published in a forthcoming paper.[7]

Figure 1. Polymerization of 1,3-cyclohexadiene to PCHD (1,2 and 1,4)

Instrumentation. The number average (Mn) and weight average (Mw) molecular weights were determined using a Perseptive Biosystems Voyager Elite DE matrix assisted laser desorption/oionization time of flight mass spectrometer (MALDI-TOF-MS) in linear mode and SEC using THF as eluent. MWD was measured by SEC using THF at 35°C. The microstructures of the 1,2/1,4 ratios were determined by ^1H and ^{13}C NMR. Characterization of the conversion process was done by IR (Bruker Vector 22), NMR (Bruker ARX 400 MHz), UV-vis (Perkin Elmer Lambda 20), and PL (Perkin Elmer LS50B). The effect on thermal stability was investigated by TGA (Mettler TG50) and DSC (Mettler DSC 30).

Conversion of PCHD derivatives to PPP. The dehydrogenation reaction involves refluxing with Chloranil and xylene as solvent.[8] A typical procedure involves the following: A mixture of 30.8g of chloranil and 125 ml of xylene, was placed in a 500 mL flask. The mixture is then heated to reflux temperatures and a mixture of 5.0 g of PCHD in 250 ml of hot xylene was added. The reflux condition is then maintained for the next 48 hours. The mixture was then cooled and filtered to obtain a brown solid product. This was then rigorously washed and extracted using a Soxhlet apparatus with the following solvent sequences: ethanol for 24 hours, ether for 8 hours, toluene for 24 hours and finally ethanol for another 24 hours. The remaining material was then oven dried and the corresponding percentage yield obtained.

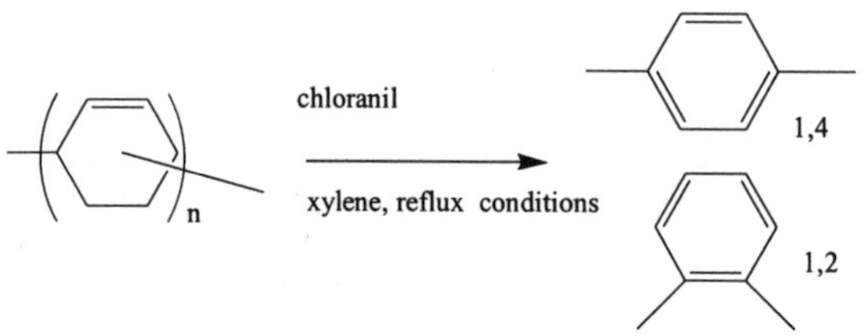

Figure 2. Conversion of PCHD to PPP using Chloranil under reflux with xylene.

RESULTS AND DISCUSSION

Polymerization of PCHD Homo- and Block copolymers. A summary of polymerization conditions and corresponding polymer product properties are shown in Table 1.

Table 1. Polymerization of 1,3-cyclohexadiene at different conditions.

Initiator System	Polymer[a]	Temp. (0C) / Solvent	Yield	Mn(K)[b]	MWD
n-BuLi/TMEDA	PCHD	40/ cyclohexane	100	27.9(30)	1.61
n-BuLi/TMEDA	PCHD	-70/ methyl-cyclohexane	100	15.1(20)	1.47
n-BuLi/TMEDA	PS/ PCHD	40/ cyclohexane		25.2(20)	1.32
s-BuLi/t-BuOK	PCHD	25/ cyclohexane	11	3.7(20)	2.21
s-BuLi/t-BuOK	PCHD	-78/ THF	87	15(20)*	1.19
s-BuLi/t-BuOK	PS/ PCHD	-78/ THF		28(30)*	1.23
benzylK	PCHD	-78/ THF	100	30(35)*	1.29
s-BuLi	PCHD	25/ benzene	63	17.3(50)	2.21
s-BuLi	PCHD	-70/ methyl cyclohexane	83	33.4(50)	2.17
s-BuLi	PCHD	-78/ THF	39		
s-BuLi	PS/ PCHD	25/ cyclohexane		Multipeaks	
CumylK / crown ether	PCHD	THF	100	19.1(20)*	1.28

a) PCHD, homopolymers; PCHD/PS block copolymers
b) from SEC (* MALDI-TOF-MS), data in parenthesis are expected MW

In general, the polymers are characterized by significant differences in MW, MWD, and various ratios of 1,2 and 1,4 on the microstructure depending on the conditions used. It has been found that lower polymerization temperatures (-10 to -70 °C) result in higher polymerization yields. Polar additives not only have a significant effect on the kinetics of the polymerization process but also change the microstructure, the ratio being determined by ^{1}H NMR and correlated with ^{13}C NMR using two-dimensional analysis techniques. For example, in hydrocarbon solvents, polymers with a high (ca 90%) 1,4 microstructure are obtained, but polymerization can take many hours to days depending on the reaction conditions. When triethylamine was used as a modifier, it increased the rate of polymerization, improving the polymer yield without having a significant impact on the microstructure. Tetramethylethylenediamine (TMEDA) causes a substantial decrease in the 1,4 content of the polymer, leading to materials with altered conformational characteristics and glass transition temperatures. Unfortunately, a gel component of extremely high MW is always a byproduct of this particular reaction condition. We are also exploring the use of other tertiary amine and ether derivatives to establish conditions, which rapidly yield polymers of the desired microstructure and narrow MWD. In summary, polar solvents tend to favor 1,2 conversion while apolar solvents favor 1,4 microstructure. Higher temperatures and higher reaction pressure conditions favor 1,2. The size of the counterion Li< Na<K, favors increasing 1,2/1,4 ratios. Side reactions which limit the obtained MW and affect the MWD usually involves reaction with Li on labile protons both with the monomer and growing polymer chains.

Polystyrene-PCHD block copolymers have been synthesized and characterized as well. The best results are obtained when the PS block is formed first, followed by the formation of the PCHD block. Initially we observed the presence of homopolymer contaminants. We were able to minimize this contamination by using hydrocarbon solvents. Recent studies of the morphology of three PS-PCHD diblocks having PCHD volume fractions around 0.37 have shown the existence of a highly unusual "cylinders in cylinders" morphology. [6] This was characterized by dispersed hexagonally packed cylinders of PCHD in a PS matrix; each PCHD cylinder has embedded in it a cylinder of PS.[9] We believe that "nanotubes" of PCHD could possibly be generated from such a system by post-morphological processing. Further investigations on this phenomena and processing is being pursued.

Conversion of PCHD to PPP. A summary of conversion and properties of the polymer on several PCHD and PS-PCHD block copolymer is outlined in Table 2.

Table 2. Summary of conversion properties of PCHD to PPP.

Polymer	UV/Vis[a] [nm]	IR [cm^{-1}][b]	NMR [ppm]	TGA [°C]	PL[c] [nm] (emission)
1,4 / 1,2 50/50 soluble in CHCl$_3$	300 to 400 shoulder	sharp, 765 and 811 broad, 2,900-3000	broad 6.6-7.5 weak at 5.5- 6.0	Steep 303-500, decline 500-800	390 (350)
1,4 / 1,2 75/25 partial in CHCl$_3$	306-440 shoulder	Weak, 765 Sharp, 810 2,900-3,000	Not very soluble		390 (350) soluble fraction
1,4 / 1,2 97/03 partial in CHCl$_3$	310-440 shoulder	Weak, 765 Sharp, 810, 2,900-3,000	Not very soluble		390 (350) soluble fraction
PS-PCHD Soluble CHCl$_3$	310-440 shoulder	Weak at 765 Sharp at 810, 2,900-3,000			390 (350)

a) UV-vis in CHCl$_3$ on soluble fraction
b) Using KBr salt on the whole product (soluble and insoluble fractions)
c) Photoluminescence on CHCl$_3$ soluble fraction

Figure 3. UV-vis spectra of PPP in KBr pellets, before conversion.

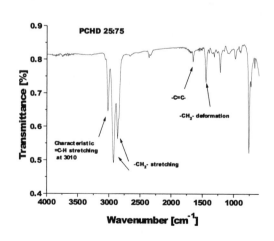

In general, the reaction of chloranil with PCHD, yields a brown powder, which have limited solubility, depending on the 1,2/1,4 ratio. From IR spectroscopy, The following peaks were observed: -CH$_2$- stretching at 2925 and 2869 cm^{-1}, aromatic C-H stretching at 3035, C=C skeletal in-plane vibrations at 1603, 1575 and 1486 cm^{-1}, and several maxima characteristic of aromatic substitution (1006,

811, 765, and 695). Depending on the 1.2/1,4 ratio, the ratio of intensity of absorption bands at 765 cm^{-1} (ortho-substitution) and at 811 (para-substitution) varies. The observed peaks at the 3000-2850 region are indicative of incomplete conversion. Some insoluble fraction was also obtained and is deemed as converted PPP (by elemental analysis). On the soluble 1,2/1,4 (50/50) derivative, we observed by NMR the formation of aromatic protons at the 8 ppm region. Analysis of the UV-vis spectra shows an absorption band that tapers to a shoulder between 300-400 nm with no significant differences between the samples. From the TGA analysis we observed that the conversion of PCHD to PPP also proceeded with heating under N$_2$ (and elemental analysis). This was observed with the almost similar trace for the converted and the unconverted polymers which differs only in the extent of slope change at the 450 °C region.

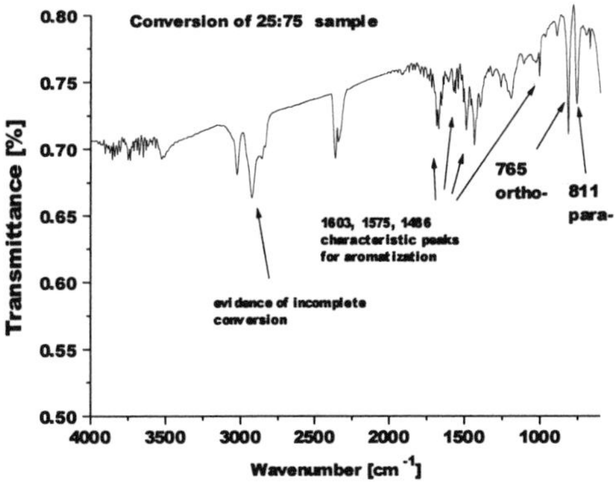

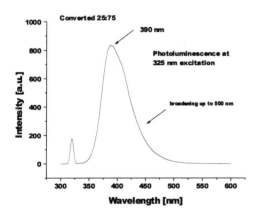

Figure 4. UV-vis spectra of PPP in KBr pellet, after conversion.

Figure 5. PL spectra of soluble PPP at 325 nm excitation wavelength.

PL spectra clearly indicate the presence of oligophenylene fluorophores on the mainchain. The observed PL for some typical poly(p-phenylene) derivatives have been observed around the 410 region.[10] This difference can indicate shorter sequences to this set of polymers or, the influence of variable 1,2 and 1,4 microstructure composition on the polymers. X-ray diffraction will be done to determine the extent of crystallinity in relation to 1,2/1,4 converted ratios. We are currently exploring the following directions: functionalization of PCHD polymers prior to conversion, other methods of dehydrogenation, and other block architectures with PCHD.

CONCLUSION

In this report, we have described our results in the anionic polymerization of 1,3 cyclohexadiene with a number of reaction conditions in order to explore controlled MW, MWD, and microstructure properties. We have found the proper combination of initiators, solvent, additives, and solvent/temp conditions to produce high MW, narrow MWD, and high 1,4 constituent polymers. We then proceeded to demonstrate the systematic conversion of these highly defined polymers into PPP derivatives. This was evident from the different spectroscopic structural properties and solubility observed depending on the microstructure of the precursor polymer. We are currently investigating the possibility of converted PPP films and relating it to the original morphology of PCHD.

ACKOWLEDGMENTS
We acknowledge support from the Army Research Office under DAAH04-94-G-0245, DAAH04-95-1-0306, and DAAG55-98-1-0005

REFERENCES

[1] M. Ivory, G. Miller, J. Sowa, L. Shacklette, R. Chance, R. Baughman,
     J. Chem. Phys. **7**, 1506 (1979).
[2] G. Grem, G. Leditzky, B. Ullrich, G. Leising, Adv. Mater. **4**, 36 (1992).
[3] G. Grem, G. Leditzky, B. Ullrich, G. Leising, Synth. Met. **51**, 389 (1992).
[4] C. Marvel, G. Hartzell, J. Am. Chem. Soc. **81**, 448 (1959).
[5] I. Natori, S. Inoue, Macromolecules **31**, 4687 (1998).
[6] J. David, S. Gido, K. Hong, J. Zhou, J. Mays submitted to Macromolecules.
[7] K. Hong, and J. Mays, manuscript in preparation
[8] G. Lefebvre, and F. Dawans, J. Polym. Sci. A: **2**, 3277 (1964).
[9] J. David, S. Gido, K. Hong, J. Zhou, J. Mays, N. Beck Tan,
     submitted to Macromolecules
[10] M. Remers, D. Neher, J. Gruner, R. Friend, G. Gelinck, J. Warman, C. Quattrocchi,
     D. dos Santos, J. Breda, Macromolecules **29**, 7432 (1996).

TEMPERATURE AND FIELD DEPENDENCE IN POLYMER LIGHT EMITTING DIODES

L. D. BOZANO*, S.A. CARTER*, J.C. SCOTT**, G.G. MALLIARAS***, P.J. BROCK**

* Physics Department, University of California, Santa Cruz, CA 95064
sacarter@cats.ucsc
** IBM Almaden Research Center, San Jose, CA 95120
*** Department of Material Science and Engineering, Cornell University, Ithaca, NY 14853

ABSTRACT

We investigate the electrical properties of Polymer Light Emitting Diodes (LED's).
The experimental data consists of steady state current-voltage characteristics and radiance as a function of temperature.
The basic LED structure is Anode/MEH-PPV (2-methoxy,5-(2'-ethyl-hexyloxy)-1,4-phenylene vinylene)/Cathode, with a polymer film 120-140 nm thick . We use different anode/cathode pairs to study transport and light emission properties. Measurements of external quantum efficiency of bipolar and monopolar devices are presented from 200 K to 300 K. The electron and hole mobilities are derived in the trap-free limit and at high voltages.

INTRODUCTION

The understanding of injection and subsequent transport phenomena of carriers in polymer light emitting diodes could be very useful to improve their performance. The operational temperature range of these devices is also of practical importance.

By changing the anode/cathode pair one can control the injection of either charge polarity. This is done by using anode and cathode materials whose work functions match either the lowest unoccupied molecular orbital or the highest occupied molecular orbital of the polymer (MEH-PPV: LUMO 3.0 eV, HOMO 5.3 eV). However several processes e.g., chemical reaction, at the polymer/electrode interface can modify the injection properties. In other words, the barrier height difference is the parameter to be considered only if no interaction occurs at the interface. For example, a device with a Sm cathode performs very poorly[1] even though Sm and Ca have similar work function. In general, electron only devices have proven to be very difficult to fabricate: for instance, traps can be formed during processing by atmospheric contamination and molecular diffusion.

The anode and cathode materials used in our experiments and their work functions are reported in Tab. I.

TAB. I: Electrodes work function.

Anode	Work function (eV)	Cathode	Work function (eV)	Device type
PAni	4.9	Ca	2.8	Bipolar
PAni	4.9	Al	4.3	Hole-dominated
TiN	3.9	Ca	2.8	Electron-dominated

Using ohmic contacts to inject the dominant carrier, the current density flowing in the single polarity trap-free device is described by the space-charge limited formula[2]

195

$$J = \frac{9}{8} \varepsilon_o \varepsilon_r \mu \frac{E^2}{L} \tag{1}$$

Where $\varepsilon_0 \varepsilon_r$ product is the permittivity of the polymer, E is the applied electric field and L is the thickness of the device; μ is the single-carrier mobility.

The analytical expression for carrier mobility[3,4] can be found using a hopping model containing two parameters, the activation energy Δ, and a temperature dependent factor γ,

$$\mu = \mu_0 \exp\left(0.89\gamma(T)\sqrt{E}\right)$$

$$\mu_0 = \mu_* \exp\left(-\frac{\Delta}{kT}\right) \tag{2}$$

Where k is Boltzman's constant, T the temperature and E the electric field, μ_0 is the so called zero-field mobility and μ_* is constant. In steady state condition, the carrier density modifies the electric field inside the device. The factor 0.89 in Equation (2) is the correction for an appropriate average of the electric field and for a non-constant mobility.[4]

Formula (1) together with (2) gives the mobility from steady state data in the trap-free limit and at moderately high electric field ($E > 10^5$ V/cm).

The external quantum efficiency η_{ext}, as the number of emitted photons exiting the device per electron injected can be written as

$$\eta_{ext} = \frac{1}{4} \frac{1}{n^2} b(N_h, N_e, \mu_h, \mu_e) \Phi_{PL}(T) \tag{3}$$

Where the 1/4 factor is the singlet contribution and $1/n^2$ is the factor representing internal trapping of the light in an anisotropic material.[5] Neither quantity depends upon temperature.
b is the recombination efficiency i.e. the percentage of electrons and holes that recombine inside the polymer. It is a function of the carrier density N, and of the hole and electron mobilities, which are temperature dependent.

Φ_{PL} , the photoluminescence efficiency, is the probability of photon emission per electron-hole recombination. Φ_{PL} is a property of the polymer and a function of temperature only.[6]

In this paper, we present temperature-dependent measurements of electron and hole mobilities derived in the trap-free limit and explain the change in quantum efficiency previously observed by us.[6]

EXPERIMENT

All devices were fabricated with the same polymer batch. The basic device has a 3-layer structure Anode/MEH-PPV/Cathode. The active area of the device is 3 mm^2 and the thickness of the polymer layer is approximately 140 nm. For the electron-dominated device a 120 nm polymer layer was spin coated on a commercially processed TiN anode. The overall thickness of a few diodes was inferred by measuring the depth of a trench cut in the MEH-PPV with an Atomic Force Microscope. The LED's were prepared in a dry nitrogen filled glove box, using techniques previously described[7]. The devices were transferred from the glove box to an Oxford MagLab cryostat. The diode current was measured by a Keithley 2400. The light output was measured by a solid state photodiode, having verified that its quantum efficiency did not change by more than 5% over the whole temperature range. The current and light output were measured every 5 K from 200 K to 300 K.[6]

RESULTS

In Fig.1 we plot the current density vs the effective electric field E, calculated as the difference $(V-V_b)/L$ where V_b is the built in potential obtained by photoconductivity measurements[8]. The data refer to one bipolar and two monopolar devices at the extreme temperatures of 200 K and 300 K. For all three devices the current density increases with the increasing temperature by approximately 2 orders of magnitudes. However electron-dominated (squares) and hole-dominated (circles) current densities are comparable at low temperature and high electric field but fairly different at higher temperature. The hole-current is larger than the electron-current by a factor of 5 at 300 K.

There is no simple model of a space charge current density for the double polarity device. The transport is complicated by recombination phenomena and E field modification. In fact, as it can be seen in Fig. 1, the double carrier device current is higher than the simple sum of the electron-dominated and hole-dominated currents.

We fit the single carrier currents in a space-charge model (for a fixed temperature) to equations (1) and (2). The data are plotted in Fig. 2, a straight line in the variables J/E^2 vs $\sqrt{E}$ on a semilog scale. Our data fit well the model for high applied electric field. From the fit we can

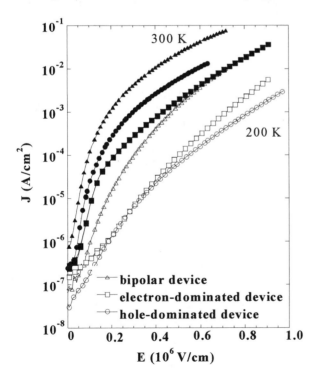

Fig.1. Current density vs electric field for the bipolar (triangles), and monopolar (square electron- dominated, circles hole-dominated) devices at the extreme temperatures of 200 K (open symbols) and 300 K (closed symbols)

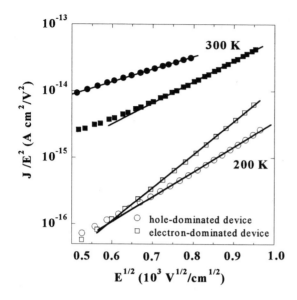

Fig.2. Electron (squares)-
and hole (circles)-dominated
devices in the space-charge
limit at 200 K (open symbols)
and 300 K (closed symbols).
The data fit the theory for
high applied electric field
(E>10⁵ V/cm)

derive the zero field mobility and the parameter $\gamma(T)$. As it can be seen in Fig 3, $\gamma(T)$ is indeed inversely proportional to the temperature and follows the empirical formula[9]

$$\gamma = \left(\frac{1}{kT} - \frac{1}{kT_0} \right) B \qquad (4)$$

Where B and T_0 are constants that can be extracted from the data, yielding the following values B=2.3 ±0.2x10⁻⁵ eV(m/V)^{1/2} and T_0=600±90 K for holes and B=2.6 ±0.4x10⁻⁵ eV(m/V)^{1/2} and T_0=880±150 K for electrons. The value of T and B for holes is consistent with previous data for another device fabricated with a slightly different PPV derivative[10].

In Fig. 3 (bottom), we plot μ_0 for the two single polarity devices as a function of 1/T to extract the activation energy Δ. We obtain 0.38±0.02 eV for holes and 0.34±0.02 eV for electrons.

Recent measurements on electrical degradation of hole dominated devices have shown that the γ factor remains substantially constant, while μ_0 decreases significantly. This change in the zero field mobility could be attributed to an increase of the number of traps.

The external quantum efficiencies are plotted in Fig. 4. Both sets of devices show an increase in quantum efficiency lowering the temperature. The change in quantum efficiency may be due to an increase of the number of carriers at lower temperatures, for constant current density. The variation is of about a factor 2 for the bipolar device and a factor of 10 for the monopolar device.

We plot in Fig. 5 the external quantum efficiency of the double polarity device vs the ratio of the hole and electrons current densities (from single polarity devices) at a fixed electric field (E= 6x10⁵ V/cm) for four different values of the temperature. The reported values are at 200 K, 230 K, 280 K and 300 K. The data show that η_{ext} improves considerably when J_e is larger than J_h[12].

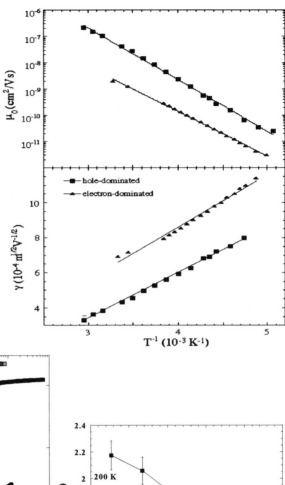

Fig.3: (top) zero-field mobility μ_0 and prefactor γ (bottom) in the electric dependent term in eq(2) vs the temperature. Electron-dominated device (triangles) and hole-dominated device (squares) show different values for these parameters.

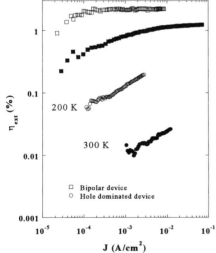

Fig.4. External quantum efficiency η_{ext} for the bipolar (squares) and hole-dominated (circles) devices

Fig. 5. External quantum efficiency for the bipolar device at 200 K, 230 K, 260 K and 300 K vs the ratio of the current densities of the 2 carriers

CONCLUSIONS

We have calculated the electron and hole mobilities from the measurement of space charge limited current density. Using the Murgatroyd's [4] approximation we obtain γ and Δ (Eq.2) for both carriers. We obtained a larger value of γ for electron than holes. This implies a stronger field dependence for electrons than for holes. Our results are consistent with previous measurements, with the exception of the value of μ_0 for electrons reported in reference[13].

The behavior of electrons vs holes explains the temperature dependence in the external quantum efficiency. We observe that a somewhat higher electron (than hole) current density improves the external quantum efficiency of the double carrier device.

ACKNOWLEDGMENTS

The authors acknowledge the support from the Packard foundation and the NSF MRSEC Center for Polymer Interfaces and Macromolecular Assembly. L. Bozano would like to acknowledge G. Alers and S. Bailard.

REFERENCES

1. I.D. Parker, J. Appl. Phys. 75, 1656 (1994).
2. M.A. Lampert and P. Mark, *Current Injection in Solids* (Academic, New York, 1970).
3. D.M. Pai, J. Chem. Phys. 52, 2285 (1970).
4. P.N. Murgatroyd, J. Phys D 3, 151 (1970).
5. N.C. Greenham, PhD Thesis, Cambridge University 1995.
6. L. Bozano, S.E. Tuttle, S.A. Carter, and P.J. Brock, Appl. Phys. Lett. 73, 3911 (1998).
7. S.A. Carter, M. Angelopoulos, S. Karg, P.J. Brock, and J.C. Scott, Appl. Phys. Lett 70, 2067 (1997).
8. G.G. Mallliaras, J.R. Salem, P.J. Brock, and J.C. Scott, J. Appl. Phys. 84, 1583 (1998).
9. W.D. Gill, J. Appl. Phys. 43, 5033 (1972).
10. P.W.M. Blom, M.J.M. de Jong, and M.G. van Munster, Phys. Rev. B. 55, 656 (1997).
11. S. Bailard private communications.
12. J.C. Scott, G.G. Malliaras, J.R. Salem,P.J. Brock, L. Bozano, and S.A. Carter in *Injection, Transport and Recombination in Organic Light-Emitting Diodes* (SPIE –Proc. 3476, San Diego, CA, 1998) pp.111-122.
13. B.K Crone, I.H. Campbell, P.S. Davis, and D.L. Smith, Appl. Phys. Lett. 73, 3162 (1998). At room temperature, for an electron dominated device (with a different anode), $\mu_0 = 5 \ 10^{-12}$ (cm^2/Vs) instead of our $2 \ 10^{-9}$ (cm^2/Vs).

SELF-ASSEMBLY PROCESSES FOR ORGANIC LED TRANSPORT LAYERS AND ELECTRODE PASSIVATION

J.E. MALINSKY*, W. LI*, Q. WANG*, J. CUI*, H. CHOU*, T.J. MARKS*, G.E. JABBOUR[†], S.E. SHAHEEN[†], J.D. ANDERSON[‡], P. LEE[‡], A.G. RICHTER[§], B. KIPPELEN[†], P. DUTTA[§], N. PEYGHAMBARIAN[†], N. ARMSTRONG[‡]
*Department of Chemistry and the Materials Research Center, Northwestern University, Evanston, IL 60208-3113, tjmarks@casbah.acns.nwu.edu
[§] Department of Physics and the Materials Research Center, Northwestern University, Evanston, IL 60208-3113
[†] Optical Science Center, University of Arizona, Tucson, AZ 85721
[‡] Department of Chemistry, University of Arizona, Tucson, AZ 85721

ABSTRACT

This contribution describes the use of self-limiting siloxane chemisoroption processes to self-assemble building blocks for the modification of vacuum-deposited organic LED (OLED) devices. One approach consists of the use of self-assembling OLED hole transport materials for application in hybrid self-assembled + vapor deposited two-layer devices. Another approach involves the application of self-limiting, chemisorptive self-assembly techniques to introduce thin dielectric films between the anode and hole transport layer of a vapor deposited two-layer OLED device.

INTRODUCTION

Impressive advances in the area of organic light-emitting diodes (OLEDs) have been seen in both vacuum-deposited small molecule[1-4] and spin-coated polymeric systems[4-7]. Basic two-layer devices consist of a hole transport layer (HTL) deposited directly onto a transparent electrode such as indium tin oxide (ITO), followed by the deposition of an emissive/electron transport layer (ETL) onto the HTL. The OLED structure is completed by the deposition of a low work function metal cathode[8-11]. Optimization of these devices requires attention to crucial materials challenges, including minimization of pinholes in the organic layers, of interlayer diffusion, of delamination of the various layers present, and minimization of thermal, chemical, and environmental degradation. In order to maximize the efficiency of these OLED devices it is also necessary to achieve a balance between the energy levels of the anode and cathode materials with the HOMO and LUMO levels, respectively, of the organic charge transporting materials.

EXPERIMENTAL

All chlorosilane reagents were synthesized, stored, and manipulated with rigorous exclusion of oxygen and moisture using Schlenk and high vacuum line techniques as well as a Vacuum Atmospheres glove box with an efficient recirculator. All solvents were dried and distilled under inert atmosphere prior to use. The synthesis of the trichlorosilane functionalized HTL building block is described elsewhere[12, 13]. All products were characterized as appropriate by ^{1}H and ^{13}C NMR, high resolution mass spectrometry, and elemental analysis. The deposition techniques for the self-assembled HTL and dielectric layers are presented elsewhere[14-16], as are the procedures for film thickness measurements performed by specular X-ray reflectivity[14-17] and scanning ellipsometry[15, 16]. Specimens were protected at all times from dust by conducting sample transfers in a class 1000 clean hood. ITO substrates were patterned and cleaned as described previously[3, 18]. The instrumentation for electroluminescence studies is described elsewhere[3].

RESULTS AND DISCUSSION

We report here two separate techniques for the modification of vacuum-deposited OLEDs. One approach consists of the deposition of a OLED hole transport layer (HTL) in which triarylamine core structures with reactive trichlorosilane functionalities undergo rapid crosslinking and densification upon spin-coating of the material in ambient atmosphere. Following a thermal cure, this process provides a robust, adherent, essentially pinhole-free charge transporting network embedded in a siloxane matrix. These HTL structures are then incorporated into a two-layer hybrid OLED by vacuum deposition of an emissive ETL [12]. The second method of OLED modification discussed here involves the use of similar self-limiting chemisorptive siloxane self-assembly techniques[13, 15, 16, 18, 19] to introduce a thin, conformal, microstructurally/electronically well-defined, nearly pinhole-free dielectric layer between the ITO anode and the HTL of a two-layer OLED device. We show that the introduction of this layer can effect significant enhancements in maximum external quantum and luminous efficiencies.

OLED Hole Transport Layers via Spin-Coating/Siloxane Condensation

In order to form a heavily crosslinked glassy HTL, a solution of trichlorosilane functionalized triarylamine **I** was spin-coated from hydrocarbon solutions in air. In the presence of moisture, **I** readily undergoes hydrolysis and subsequent ≡Si-OH + HO-Si≡ condensation[20-22] which, upon thermal curing (120 °C in air), produces a smooth, hard, thermally stable, and adherent, glassy matrix (**II**) as seen in Figure 1. By varying the concentrations of the spin-coating solutions, the thickness of the films can be controlled (100-700 Å) as assayed by ellipsometry on duplicate samples cast on (100) Si wafers. AFM studies of these spin-coated films on clean ITO (rms roughness ~30 Å) show smooth (rms roughness ~15 Å), conformal films with no obvious cracks or pinholes. The reduction in the roughness of the substrate after deposition of the HTL suggests that the spin-coating of the cross-linked **II**-HTL layer effects planarization of the surface.

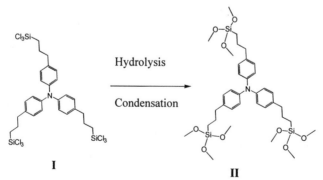

Figure 1. Conversion of trichlorosilane-functionalized triarylamine building block **I** to a siloxane crosslinked hole transport layer (HTL) **II**.

In order to study the extent to which the HTL precursor **I** undergoes hydrolysis, a cast film was mechanically detached from a glass substrate. Elemental analysis of this removed

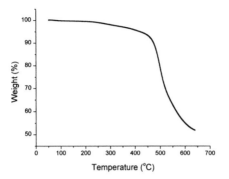

Figure 2. Thermogravimetric analysis under Ar of a **II** sample obtained from a spin-coated, cured film. The temperature ramp is 15°C/min

material revealed Cl levels below the detection limit, suggesting essentially complete hydrolysis of the precursor. Thermogravimetric analysis of this material as shown in Figure 2 reveals only ~2% weight loss in the 50-300 °C range suggesting that there are few uncondensed Si-OH functionalities. A ~5% weight loss is observed in the 300-450 °C range, demonstrating considerable resistance to thermal breakdown. DSC analysis of this material shows no thermal transitions in the 50-400°C range, implying a high degree of condensation/densification.

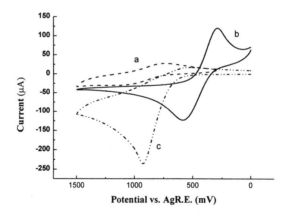

Figure 3. Cyclic voltamograms of a) ITO/**II**(550 Å) with 0.1 M TBAHFP electrolyte in 1:1 benzene:acetonitrile, b) bare ITO in contact with a ferrocene solution (10^{-3}M) in 1:1 benzene:acetonitrile containing 0.1 M TBAHFP electrolyte, c) ITO/**II** (550 Å) with a ferrocene solution (10^{-3}M) in 1:1 benzene:acetonitrile containing 0.1 M TBAHFP electrolyte.

Cyclic voltammetry of cured, **II**-based films cast on ITO electrodes reveals them to be electroactive, virtually pinhole-free, and capable of efficient hole transport. Oxidation of the triarylamine groups occurs ca. 450 mV positive of ferrocene (~950 mV vs. AgRE) of ferrocene as seen in Figure 3a. The ratio of oxidative to reductive peak currents indicates that HTL oxidation of the triarylamine groups present in the film is reversible. The ~250 mV separation between oxidative and reductive peak potentials suggests a kinetically inhibited oxidation process. A redox-active film which supports rapid counterion penetration/ejection during redox cycling is expected to exhibit voltametric responses with well-defined gaussian shapes centered at the formal potential for the redox process. Counterion mobility during redox cycling of the **II**-triarylamine groups appears to be inhibited, as indicated by the peak shapes. This is likely due to the highly crosslinked microstructure[23]. Figure 3b shows the cyclic voltammogram of a bare ITO electrode in a 10^{-3} ferrocene solution, which may be compared to Figure 3c, which shows the cyclic voltammogram of a **II**-coated ITO electrode immersed in a similar solution. The lack of significant current flow near the formal oxidation potential of ferrocene in the latter indicates an inhibition toward ferrocene oxidation. This would be as expected for a pinhole-free surface coverage. As the triarylamine oxidation potentials are reached, an electrocatalytic current response is observed whereby ferrocene is oxidized at a diffusion-controlled rate. This result indicates facile oxidation of the triarylamine group at the ITO/HTL interface (hole injection), and rapid hole migration through the film such that ferrocene oxidation occurs at the film-solution interface.

Figure 4 shows an in situ optical absorption spectrum of an ITO/**II** film immersed in the same electrolyte solution as Figure 3c under zero bias as well as at a potential of ~1.1 V vs. AgRE. The oxidation of the HTL is accompanied by the appearance of a new absorption band at ~ 690 nm, with a shoulder at ~600 nm. These features can be assigned to triarylamine radical cations by the similarity to recent solution phase spectroelectrochemical data for triarylamine oxidation in acetonitrile[24], and triarylamine radical cation absorptions reported at ~560 and 650 nm in 77K s-BuCl matrices[25]. In regard to the latter data, it is reasonable that radical cation spectral shifts of this magnitude would be observed between polar and non-polar media. Bässler and coworkers recently reported similar observations during the forward bias operation of an OLED with a triarylamine-based HTL[26].

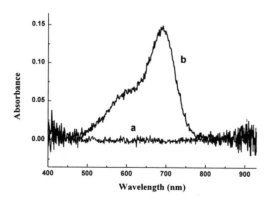

Figure 4. Spectroelectrochemistry of ITO/**II** (550 Å) under (a) zero bias and (b) under a forward bias (i.e., positive on the ITO electrode) of 1074 mV (vs. AgRE).

Films of **II** were incorporated into hybrid OLED devices in which varying thicknesses of **II** were deposited onto ITO, followed by a conventional vapor deposited aluminum quinoxolate (Alq) electron transport/emissive layer, and finally by a Mg cathode. The devices had the structure: ITO/**II** variable thickness/Alq(500 Å)/Mg (1500 Å), and were characterized in terms of luminance characteristics as described elsewhere[3, 13]. Figure 5 shows the device characteristics of four hybrid LEDs assembled in this manner as a function of HTL thickness. All samples showed good performance, with turn-on voltages of ~6V and maximum forward light outputs of ~1,500 cd/m². Devices with thinner HTLs showed slightly lower turn-on voltages. At low voltages, no leakage current was observed demonstrating the robust and pinhole-free properties of these films. There is no obvious relationship between **II** film thickness and external quantum efficiency, with all four devices exhibiting efficiencies of ~0.2%.

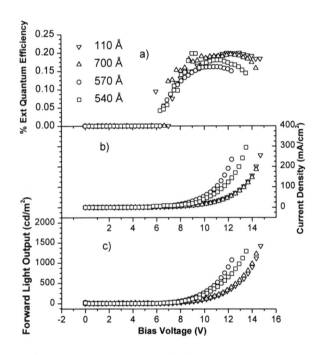

Figure 5. Device characteristics of a hybrid OLED fabricated from ITO/**II**/ 500 Å AlQ (vacuum deposited)/1500 Å Mg. a) external quantum efficiency, b) I-V characteristics, c) forward light output. The thickness of the HTL is: 110 Å, 540 Å, 570 Å, and 700 Å.

Organic LED Electrode Passivation and Charge Injection Balance

As shown in Figure 6, layers of the octachlorotrisiloxane capping agent[15, 16] can be sequentially chemisorbed from dry hydrocarbon solutions onto clean ITO-coated glass. Subsequent hydrolysis of the deposited monolayer followed by thermal curing/crosslinking in air yields a thin dielectric layer on the surface of the ITO. Repetition of this process allows the deposition of dielectric films, the total thicknesses of which are determined by the number of times that the process is repeated. Specular x-ray reflectivity measurements on coated Si(111) substrates indicate that the total dielectric film thickness increases monotonically with repeated layer deposition. Each deposition applies ~8.3 Å of dielectric material on the Si substrate. It was determined that the electron density of the self-assembled dielectric material, as seen in the x-ray reflectivity measurements, is ~85% of that of the native SiO_2 on the crystalline substrate.

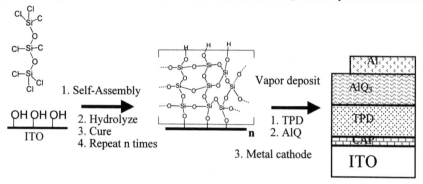

Figure 6. Procedure for the layer-by-layer chemisorptive self-assembly of siloxane dielectric "capping" layers. The thickness of each layer applied on each successive deposition is ~8.3 Å

Cyclic voltammograms of a millimolar ferrocene solution using ITO electrodes that were coated with various thicknesses of self-assembled dielectric material are shown in Figure 7. As judged by the separation between anodic and cathodic peaks, and by the overall magnitude of the current flowing at any potential, there is a successive passivation of the ITO surface toward this redox chemistry as the capping layer thickness is increased. Compared with electrochemical responses seen on other passivated electrode surfaces, where pinholes with areas totaling less than 1% of the geometric area lead to current/voltage responses similar to those in Figure 7[27], we believe that 3-4 layers of dielectric capping layer are sufficient to eliminate the majority of pinholes which would compromise OLED device performance.

Vacuum deposited OLED devices[3] consisting of a TPD hole transport layer, an AlQ electron transport/emissive layer, and an Al cathode were fabricated using ITO substrates which had been modified with various thicknesses of the self-assembled dielectric material. The devices thus have the structure: ITO/ SA capping layer/TPD (600 Å)/ AlQ (600 Å)/Al (2000 Å). Figure 8 shows the external quantum efficiency (photons/electron), current density, and forward light output characteristics for a series of substrates with dielectric thicknesses ranging from 0-33 Å. It can be seen that the effect of 8 Å of dielectric material is to substantially increase the operating voltage, decrease the light output at a given voltage, and decrease leakage currents, while more than doubling the maximum quantum efficiency of the device. As the thickness of the dielectric layer increases, the operating voltage decreases to values nearing the bare electrode values while maintaining high efficiency values. As the dielectric material becomes even thicker, the operating voltage eventually begins to increase again and the quantum efficiency falls. These data tend to suggest that in these modified ITO devices, anode hole injection is more efficient than electron injection from the cathode. Such an imbalance would give rise to inefficient carrier recombination, and in device regions not conducive to high luminance efficiencies (e.g., near the cathode). It appears that the ITO functionalization tends to correct the hole/electron imbalance. Thus, as dielectric layers are chemisorbed, the hole injection imbalance is decreased and the luminous recombination occurs in a more favorable region of the device. As the dielectric layer thickness reaches ~33 Å, hole injection by tunneling becomes inefficient versus electron injection, and light output, current density, and external quantum efficiency for a given voltage fall, as would be appropriate for this model. The intermediate regime of a single self-assembled dielectric is intriguing and appears to reflect a complex interplay of the above effects combined with modification of ITO surface states and/or confinement effects in the thin capping layer.

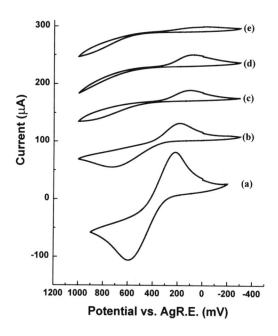

Figure 7. Cyclic voltammetry of a 10^{-3} M ferrocene solution in acetonitrile, at a) a bare ITO surface, b) one layer, c) two layers, d) three layers, e) four layers of the self-assembled dielectric film. The sweep rate is 100 mV/sec, and the electrode area, ~0.7 cm^2

CONCLUSIONS

The results of this study show that a spin-coating/siloxane cross-linking approach to OLED charge transport layers provides a high throughput method to achieve robust, pinhole-free, adherent thin films with covalently interlinked, glassy structures of readily controllable thickness. While this study has focused on the hole transport layer of OLEDs, the extension of this technique to emissive and electron transport agents is also possible. Similar techniques have also been successfully applied to the deposition of thin dielectric layers between the anode and hole transport materials of OLEDs in order to effect an increase in the external quantum efficiency of the devices.

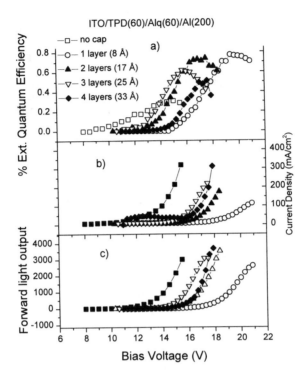

Figure 8. External quantum efficiency (a), current density (b), and forward light output (c) vs. voltage plots for ITO/capping layer/TPD/AlQ/Al OLEDs showing the effect of various thicknesses of self-assembled dielectric material (capping layer) between the anode and the hole transport layer.

ACKNOWLEDGEMENTS

We are grateful to the ONR through the Center for Advanced Multifunctional Nonlinear Optical Polymers and Molecular Assemblies (CAMP) MURI (N00014-95-1-1319) and to the NSF-MRSEC program through the Northwestern Materials Research Center (DMR-9632472) for support of this research. We thank Prof. Mark Ratner for stimulating discussions.

REFERENCES:

[1]   T. Tetsui, *MRS Bulletin*, pp. 31 (1997).
[2]   J. R. Sheats, H. Antoniadis, M. Hueschen, W. Leonard, J. Miller, R. Moon, D. Roitman,
       A. Stocking, *Science* **1996**, *273*, 884.
[3]   G. E. Jabbour, Y. Kawabe, S. E. Shaheen, J. F. Wang, M. M. Morrell, B. Kippelen, N.
       Peyghambarian, *Appl. Phys. Lett.* **1997**, *71*, 1762.
[4]   L. J. Rothberg, A. J. Lovinger, *J. Mater. Res* **1996**, *11*, 3174.
[5]   A. Kraft, A. C. Grimsdale, A. B. Holmes, *Angew. Chem. Int. Ed.* **1998**, *37*, 402.
[6]   O. Onitsuka, A. C. Fou, M. Ferreira, B. R. Hsieh, M. F. Rubner, *J. Appl. Phys.* **1996**, *80*,
       4067.
[7]   N. C. Greenham, R. H. Friend, *Solid State Phys.* **1995**, *49*, 1.
[8]   Y. Hamada, H. Kanno, T. Sano, H. Fujii, Y. Nishio, H. Takahashi, T. Usuki, K. Shibata,
       Appl. Phys. Lett. **1998**, *72*, 1939.
[9]   E. M. Han, L. M. Do, N. Yamamoto, M. Fujihira, *Chem. Lett.* **1995**, 57.
[10]   C. Adachi, K. Nagai, N. Tamoto, *Appl. Phys. Lett.* **1995**, *66*, 2679.
[11]   F. Li, H. Tang, J. Anderegg, J. Shinar, *Appl. Phys. Lett.* **1997**, *70*, 1233.
[12]   W. Li, Q. Wang, J. Cui, H. Chou, S. E. Shaheen, G. E. Jabbour, J. Anderson, P. Lee, B.
       Kippelen, N. Peyghambarian, N. R. Armstrong, T. J. Marks, *Adv. Mater.* in press.
[13]   J. E. Malinsky, W. Li, H. Chou, W. Ma, L. Geng, T. J. Marks, G. E. Jabbour, S. E.
       Shaheen, B. Kippelen, N. Peyghambarian, P. Dutta, A. G. Richter, N. R. Armstrong, P.
       A. Lee, J. D. Anderson, *SPIE* **1998**, *3281*, 148.
[14]   S. B. Roscoe, A. K. Kakkar, T. J. Marks, A. Malik, M. K. Durbin, W. P. Lin, G. K.
       Wong, P. Dutta, *Langmuir* **1996**, *12*, 4218.
[15]   W. Lin, W. P. Lin, G. K. Wong, T. J. Marks, *J. Amer. Chem. Soc.* **1996**, *118*, 8034.
[16]   W. Lin, T. L. Lee, P. F. Lyman, J. J. Lee, M. J. Bedzyk, T. J. Marks, *J. Amer. Chem. Soc.*
       1997, *119*, 2205.
[17]   A. Malik, W. Lin, M. K. Durbin, T. J. Marks, P. Dutta, *J. Chem. Phys.* **1997**, *107*, 645.
[18]   J. E. Malinsky, G. E. Jabbour, S. E. Shaheen, J. D. Anderson, A. G. Richter, T. J. Marks,
       N. R. Armstrong, B. Kippelen, P. Dutta, N. Peyghambarian, *Adv. Mater.* **1999**, *11*, 227.
[19]   J. W. Klaus, O. Sneh, S. M. George, *Science* **1997**, *278*, 1934.
[20]   U. Schubert, N. Hüsing, A. Lorentz, *Chem. Mater.* **1995**, *7*, 2010.
[21]   B. Hardman, A. Torkelson, *Encyclopedia of Polymer Science and Technology, Vol. 15*,
       Wiley, New York **1989**;p.265.
[22]   R. J. P. Corriu, D. Leclercq, *Angew. Chem. Int. Ed.* **1996**, *35*, 1420.
[23]   P. E. Smolenyak, E. J. Osburn, S. Y. Chen, L. K. Chau, D. F. Obrien, N. R. Armstrong,
       Langmuir **1997**, *13*, 6568.
[24]   J. Anderson, J.-P. Dodelet, S. Barlow, S. Thayumanuvan, S. Marder, N. R. Armstrong, .
[25]   T. Shida, *Electronic Absorption Spectra of Radical Ions*, Elsevier, New York **1988**;p.211.
[26]   M. Redecker, H. Bässler, H. H. Hörhold, *J. Phys. Chem* **1997**, *101*, 7398.
[27]   H. Finklea, in *Electroanalytical Chemistry, Vol. 19*,(Eds: A. J. Bard, I. Rubenstein),
       Marcel Dekker, New York **1996**, p. 177.

1.5 µm LUMINESCENCE FROM ErQ BASED ORGANIC LIGHT EMITTING DIODES

R.J. CURRY, W.P. GILLIN

Department of Physics, Queen Mary and Westfield College, University of London, London, E1 4NS, United Kingdom.

ABSTRACT

Organic light emitting diodes have been fabricated using erbium *tris*(8-hydroxy-quinoline) as the emitting layer and N, N'-diphenyl-N,N'-bis(3-methylphenyl)-1,1'-biphenyl-4,4'-diamine (TPD) as the hole transporting layer. Room temperature electroluminescence was observed at 1.5 µm due to intra-atomic transitions between the $^4I_{13/2}$ and $^4I_{15/2}$ levels in the Er^{3+} ion. These results make the possibility of producing silicon compatible 1.5 µm technology based on such devices a reality.

INTRODUCTION

Erbium doped materials have for many years been the subject of much interest due to their applications in optical fibre communications. The Er^{3+} ion has a sharp luminescence centred at 1.5 µm due to an intra-4 f shell transition between the first excited state ($^4I_{13/2}$) and the ground state ($^4I_{15/2}$). Given that silicon is transparent at 1.5 µm it has been doped with erbium, often with other dopants such as oxygen and fluorine, with the hope of producing a silicon based 1.5 µm emitter technology [1]. There are however, problems with this approach as the erbium related luminescence tends to quench at room temperature.

Organic light emitting diodes (OLED's) have been the subject of much research since Tang and VanSlyke demonstrated electroluminescence from aluminium *tris*(8-hydroxyquinoline) (AlQ) [2]. Since then considerable work has been done on improving OLED characteristics such as lifetime, brightness and efficiency. In this letter we report the fabrication of an OLED emitting electroluminescence at 1.5 µm using erbium *tris*(8-hydroxyquinoline) (ErQ) as the emitting layer and N,N'-diphenyl-N,N'-bis(3-methylphenyl)-1,1'-biphenyl-4,4'-diamine (TPD) as the hole transporting layer.

EXPERIMENT

Photoluminescence was excited using the 351 nm line from an argon ion laser. Room temperature photoluminescence was recorded for both the 'band-edge' luminescence, using a 0.5 µm Blaze grating and a S-20 photomultiplier, and for the 1.5µm erbium related luminescence, using a 1µm Blaze grating and a liquid nitrogen cooled Ge detector. The resolution of the system was 0.8 nm for all spectra. OLED structures of area 2 mm x 2 mm were produced by evaporating 50 nm of TPD on to an indium tin oxide coated glass substrate, with a sheet resistance of 20 $\Omega/\square$, followed by 60 nm of ErQ. 220 nm of aluminium was then evaporated to form the top electrode. Electroluminescence spectra were recorded by the same equipment used for the photoluminescence.

RESULTS AND DISCUSSION

Figure 1 shows the "band edge" electroluminescence emitted from a TPD/ErQ diode, under a driving voltage of 25 V, along with the photoluminescence recorded from the bulk ErQ. It can be seen that there is little difference between the two spectra, although the 'band-edge' electroluminescence is approximately two orders of magnitude weaker than the photoluminescence and shifted slightly to lower energies, from 2.08 meV to 2.06 meV. It can also be seen in figure 1 that the photoluminescence spectra has some sharp structures on the low energy tail which do not appear to be present in the electroluminescence. It should be noted that the photoluminescence spectra presented in figure 1 does not have the high energy component observed in our original material [3] and this is probably due to the improved purity of the material used in this study. Both the electroluminescence and photoluminescence spectra peak at ~600 nm and have a full width at half maximum (FWHM) of ~400 meV. It must be stressed that the 'band-edge' electroluminescence of ErQ based diodes is considerably weaker (>10⁴) in intensity than the electroluminescence, observed under the same operating conditions, from OLED's with other group III chelates (such as AlQ) used as the emitting layer.

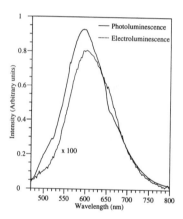

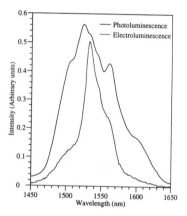

Figure 1. The 300K 'band-edge' photoluminescence of ErQ excited using the 457 nm line from an argon ion laser and the electroluminescence from an ITO/TPD/ErQ/Al OLED.

Figure 2. The 300K erbium related photoluminescence excited using the 457 nm line from an argon ion laser and the erbium related electro-luminescence from an ITO/TPD/ ErQ/Al OLED.

Figure 2 shows the erbium related electroluminescence, obtained under the same conditions as the 'band edge' electroluminescence, along with the photoluminescence for the same region. The two spectra are markedly different with the electroluminescence having a FWHM of ~31 nm compared to ~84 nm for the photoluminescence. For the photoluminescence spectra there are peaks visible at 1506, 1524, 1532, 1544, 1560 and 1600 nm. With the brightest emission peak being at 1524 nm. For the electroluminescence however, the 1524 nm peak has been virtually eliminated and the 1560 nm peak very much reduced. In addition the intensity of the 1506 nm and 1600 nm peaks are much lower and it is this in particular which results in the much narrower FWHM.

These differences between the photoluminescence and the electroluminescence may be indicative of different forms of ErQ. It is known that the Stark splitting of the energy levels in Er^{3+} depends on the local electric field seen by the atom [4]. Furthermore it is known that for AlQ there are two isomers of the molecule, the *fac* (C_{3v} symmetry) and *mer* (C_{2v} symmetry) [5]. It is possible that the material used for the photoluminescence, which has not been sublimed, could be a mixture of these two isomers resulting in emission from erbium atoms in different local environments. However, the sublimation required to form the OLED may have resulted in this material having a higher concentration of one of the isomers.

We have used a frequency resolved spectroscopy approach [6] to measure the lifetime of the erbium related luminescence as a function of the emission wavelength. If the photoluminescence did contain emission from two different isomers and the erbium in these isomers had different lifetimes this approach would have allowed us to differentiate the two components. However, the experiment showed that even if there are two different isomers responsible for the broad photoluminescence spectra, they both have an identical recombination lifetime of ~100μs.

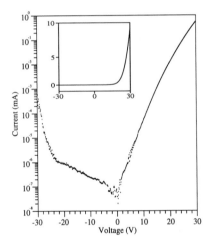

Figure 3. The current-voltage characteristic of a 4 mm² ITO/TPD/ErQ/Al OLED.

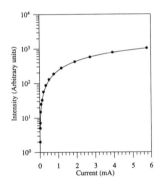

Figure 4. The 1.5 μm electroluminescence intensity against drive current for a 4 mm² ITO/TPD/ErQ/Al OLED.

Figure 3 shows the current-voltage characteristic for a typical device, and figure 4 shows the erbium related 1.5 μm electroluminescence intensity as a function of driving current. From the I-V characteristics it can be seen that the device turn on voltage is ~12 V. This high turn on voltage is indicative of poor injection efficiencies and is probably caused by the use of aluminium as the contact electrode. There is no reason to suppose that this turn on voltage can not be reduced as has been achieved with AlQ based diodes. The device shows excellent diode characteristics however, and does not start to break down until reverse biases as high as -25 V (figure 3). We have performed some preliminary measurements of electroluminescence intensity against drive current. These show the electroluminescent intensity rising exponentially towards a maximum value. The solid line in figure 4 is a fit to the data which shows that the intensity is tending to a value roughly twice that obtained at 30V. Given the long lifetime for radiative recombination of the excited erbium we would expect to see the intensity tend towards a saturated value in this manner.

It has been demonstrated by Zhou *et al* [7] and Kim *et al* [8] that OLED's can be fabricated using *p*-type doped silicon as the anode with TPD and AlQ as the organic layers. Given the results presented here it is now feasible that a 1.5 μm silicon integrated emitter technology can be produced, and furthermore it is probable that a population inversion of the erbium ions may be possible. Therefore by designing an appropriate cavity it may well be possible to achieve a 1.5 μm laser based on silicon.

CONCLUSIONS

In conclusion, we have demonstrated 1.5μm room temperature electroluminescence from an organic light emitting diode based on ITO and using TPD as a hole transport layer and ErQ as the emitting layer.

ACKNOWLEDGMENTS

The authors would like to thank M. Somerton for his help in producing and purifying the ErQ.

REFERENCES

1. A. Polman, J. Appl. Phys., **82**(1), 1, 1997.

2. C.W. Tang and S.A. VanSlyke, Appl. Phys. Lett., **51**(12), 913, 1987.

3. W.P. Gillin and R.J. Curry, Appl. Phys. Lett., **74**(6), 798, 1999.

4. Y.S. Tang, K.C. Heasman, W.P. Gillin and B.J. Sealy, Appl. Phys. Lett., **55**(5), 432, 1989.

5. K. Sano, Y. Kawata, T.I. Urano and Y. Mori, J. Mater. Chem., **2**(7), 767, 1992

6). S.P. Depinna and D.J. Dunstan, Phil. Mag. B, **50**(5), 579, 1984.

7). X. Zhou, J. He, L.S. Liao, M. Lu, Z.H. Xiong, X.M. Ding, X.Y. Hou, F.G. Tao, C.E. Zhou and S.T. Lee, Appl. Phys. Lett., **74**(4), 609,1999.

8). H.H. Kim, T.M. Miller, E.H. Westerwick, Y.O. Kim, E.W. Kwock, M.D. Morris and M. Cerullo, J. Lightwave Technol., **12**(12), 2107, 1994.

FREQUENCY RESOLVED PHOTOLUMINESCENCE (PL), DELAYED FLUORESCENCE AND TRIPLET-TRIPLET ANNIHILATION IN π-CONJUGATED POLYMERS

J. Partee,[1,2] E. L. Frankevich,[1,3,4] B. Uhlhorn,[1,2] J. Shinar,[1,2,3] Y. Ding,[5] and T. J. Barton[5]
[1]*Ames Laboratory – USDOE,* [2]*Physics and Astronomy Department, and*
[3]*International Institute of Theoretical and Applied Physics,*
Iowa State University, Ames, IA 50011
[4]*Institute of Energy Problems of Chemical Physics, Moscow 334, Russia*
[5]*Ames Laboratory – USDOE and Chemistry Department,Iowa State University, Ames, IA 50011*

ABSTRACT

The delayed fluorescence of poly(*p*-phenylene vinylene) (PPV) and poly(*p*-phenylene ethynylene) (PPE) derivative solids and frozen solutions at 20 K is described. It provides strong evidence for triplet-triplet annihilation to singlets excitons accounting for up to ~3% of the total emission in PPV films and ~1.5% in PPE powder. It also yields triplet lifetimes of 70 and 110 μs in PPV films and frozen solutions, and ~200 and ~500 μs in PPE powder and frozen solutions, respectively.

INTRODUCTION

Very extensive studies of the photoluminescence (PL) of π-conjugated polymer films and the electroluminescence (EL) of polymer light-emitting devices (LEDs) have been reported during the past fifteen years [1]. Yet several basic issues have remained unanswered. In particular, no quantitative observations of long-lived (microseconds to milliseconds) emission due to delayed fluorescence (DF) have been reported. In contrast, in small π-conjugated molecules such DF due to triplet-triplet (*T-T*) exciton annihilation to singlet excitons

$$^3T + {}^3T \rightarrow {}^1S^*$$ (1)

which relax to the lowest (luminescent) 1^1B_u state was identified unambiguously and studied in detail [2]. On a broader level, understanding triplet dynamics is important for small organic molecular and polymer LEDs, as they may be the most prolific of states generated by fusion of the positive and negative polaronic carriers. Indeed, the 25% theoretical upper limit of the EL yield of such LEDs is derived from spin statistics, which should yield nonluminescent triplets in 75% of these fusion events. Yet this argument is obviously simplistic, and an established picture of the nature of the coupling between the singlet excitons, triplet excitons, and the intermolecular or interchain polaron pairs may lead to significant modifications of the simple spin-statistical result.

The lowest energy level of the interchain polaron pairs is widely believed to be slightly (~0.1 eV) lower than that of the 1^1B_u [3] and consequently decidedly higher than that of the lowest intrachain 1^3B_u triplet [1,2,4]. Some studies have suggested that under steady-state photoexcitation the population of the triplet (parallel spin) polaron pairs is greater than that of the singlet (antiparallel spin) pairs, which is depleted by nonradiative recombination to the ground state [5]. Other studies, however, have suggested that the triplet pair population, which may recombine to intrachain triplet excitons, is lower than that of the singlet pair population [6]. These latter studies also predicted that in addition to direct nonradiative decay to the ground state, the intrachain triplets decay by *T-T* annihilation to singlets (Eq. (1)) resulting in DF.

Mat. Res. Soc. Symp. Proc. Vol. 561 © 1999 Materials Research Society

The unmistakable presence of intrachain triplet excitons in π-conjugated polymers has been observed by optically detected magnetic resonance (ODMR) [6-11] and their dynamics were studied by this technique and photoinduced absorption (PA) [4,9,12]. Indeed, the numerous cw and time-resolved PA studies showed that the polarons and triplet excitons are the only significant long-lived photoexcitations in such polymers [4,9,12]. Hence T-T annihilation would be the only plausible source of any DF. Yet no direct evidence for such DF has been reported. This paper provides such evidence by describing the PL response to modulated excitation of 2,5-dihexoxy poly(p-phenylene vinylene) (DHO-PPV) powder, 2,5-dioctoxy PPV (DOO-PPV) film and frozen toluene solution, and 2,5-dibutoxy poly(p-phenylene-ethynylene) (DBO-PPE) powder and frozen toluene solution. Both PPVs and PPEs are strongly luminescent materials [10,11,14, 15]. The latters' gap is ~2.6 eV [10,11,13] and they have been used as the emitting layer in polymer LEDs [11,13-15]. The present results show that bimolecular recombination, due ostensibly to T-T annihilation to singlets, is the source of DF which contributes up to ~3% of the total emission in DOO-PPV film, ~1.5% in DBO-PPE powder, and ~2% in frozen dilute DBO-PPE/toluene solution at 20 K at an absorption rate of ~10^{22} photons/cm^{3}s. The measurements also yield triplet lifetimes of 70 and 110 μs in the PPV solids and frozen solutions, and ~200 and ~500 μs in PPE powder and frozen solution, respectively.

EXPERIMENTAL PROCEDURE

Powder and free-standing films of the PPV and PPE derivatives were vacuum sealed in quartz tubes. Solution samples were frozen, degassed, and then vacuum sealed. The PL system, which is part of an ODMR spectrometer, was described previously [5,7,8]. In this work, however, the 488 nm Ar$^+$ laser excitation power $5 \leq P \leq 35$ mW was modulated at frequencies 10 Hz $\leq v = \omega/2\pi \leq 10$ kHz by the Pockels cell of a Cambridge Instruments Model LS100 laser stabilizer, so the photon absorption rate $G(t) = \alpha P(t)$ was given by

$$G(t) = G_0\left(1 + a\sin\omega t\right) \qquad (2)$$

The PL was detected by a fast Hamamatsu photomultiplier tube, and the in-phase x and quadrature y Fourier components of the incident laser signal and the PL at $v = \omega/2\pi$ and $2v$ were monitored by an EG&G 5210 lockin amplifier.

The intrachain triplet exciton population n_T is governed by the rate equation [6]

$$\frac{dn_T}{dt} = G(t)\beta_T - k_T n_T - \gamma n_T^2 \qquad (3)$$

where β_T is the yield of intrachain triplet excitons, k_T is their monomolecular decay rate, and γn_T^2 is the rate of triplet-triplet annihilation. At low excitation intensity γn_T^2 can be neglected, and assuming a solution of the form

$$n_T(t) = A[1+B\sin(\omega t+\varphi)] \qquad (4)$$

and substitution of this expression and Eq. (2) in Eq. (3) yields

$$n_T(t) = G_0\beta_T\tau_T\left[1+\frac{a}{\sqrt{1+\omega^2\tau_T^2}}\sin(\omega t + \varphi)\right] \qquad (5)$$

where $\tau_T = 1/k_T$ is the monomolecular triplet decay time and the phase shift $\varphi = \arctan(-\omega\tau_T)$. The intensity per unit volume of the DF I_{DF} is given by [6]

$$I_{DF} = \beta_S \frac{\gamma}{9} n_T^2 \qquad (6)$$

where β_S is the prompt fluorescence (PF) yield and the spin statistical factor of 1/9 is the fraction of pairs of triplets which are in the singlet configuration. Substitution of Eq. (5) in Eq. (6) yields a component $I_{DF,2\omega}$ of the DF which oscillates at 2ω:

$$I_{DF,2\omega} = \frac{\gamma}{9} G_0^2 \tau_T^2 a^2 \beta_T^2 \beta_S \frac{1}{1+\omega^2\tau_T^2} \sin(2\omega t + 2\varphi - \frac{\pi}{2}) \qquad (7)$$

Since the component $I_{PF,\omega}$ of the PF at ω is given by

$$I_{PF,\omega} = G_0 \beta_S a \sin(\omega t) \qquad (8)$$

the ratio of the amplitude of $I_{DF,2\omega}$ to the amplitude of $I_{PF,\omega}$ is

$$R_{PL} = \frac{1}{9} \frac{\gamma G_0 a \beta_T^2 \tau_T^2}{1+\omega^2\tau_T^2} \qquad (9)$$

Hence, a measurement of R_{PL} vs ω yields τ_T. In addition, the fraction I_{DF}/I_{PF} of the total emission due to DF resulting from T-T annihilation under cw conditions is then equal to the value of R_{PL}/a at low ω where $\omega \tau_T \ll 1$.

To correct for the system response and any Fourier component of the modulated laser beam at 2ω, the ratios of the amplitudes of the x and y components of the laser beam signal at 2ω to the modulus at ω ($R_{Las,x}$ and $R_{Las,y}$, respectively) were subtracted from the similarly defined $R_{PL,x}$ and $R_{PL,y}$, respectively. The value of R_{PL}, corrected for the system response, is then

$$R'_{PL} = \sqrt{\left(R_{PL,x} - R_{Las,x}\right)^2 + \left(R_{PL,y} - R_{Las,y}\right)^2} \qquad (10)$$

RESULTS AND DISCUSSION

Figure 1 shows R'_{PL} vs $\nu = \omega/2\pi$ of (i) aged DHO-PPV powder [16], (ii) fresh DOO-PPV film [17] and (iii) <0.025 mg/ml toluene solution of that DOO-PPV at 20 K. The phases of the laser and PL signals at 2ω confirmed that the PL signal at 2ω is positive (see Eq. (7)) and indeed results from DF due to a bimolecular process. We also note that the 3ω component of the PL was negligible. The excellent agreement between the results shown in Fig. 1 and a Lorentzian $A/[1 + \omega^2\tau^2] + C$ (allowing for an additive constant C) yields lifetimes $\tau = 70$ μs in the DHO-PPV powder and DOO-PPV film and $\tau = 110$ μs in the frozen DOO-PPV:toluene solutions.

The inset of Figure 1 shows I_{DF}/I_{PF} vs the laser power P. The solid lines are the behavior predicted by the relation

$$\frac{I_{DF}}{I_{PF}} = c_1 \frac{\left(\sqrt{1+c_2 P} - 1\right)^2}{P} \qquad (11)$$

where $c_1 = k_T^2/36\gamma$ and $c_2 P = 4\gamma G_0 a \beta_T/\alpha k_T^2$ (see Eq. (12) in ref. 6). The observed saturation of I_{DF}/I_{PF} at $P \geq 10$ mW is due to the significant contribution of the γn_T^2 term in Eq. (3) in lowering the triplet population at these excitation intensities. It can be shown that in the DHO-PPV powder and DOO-PPV film at $P = 10$ mW the contribution of the γn_T^2 term shortens the overall lifetime of the triplets by ~20% [18]. Hence in that case the true monomolecular lifetime is ~90 μsec rather than ~70 μsec. Note that at $P = 35$ mW the DF contributes ~3% and ~1% of the total emission in the DOO-PPV film and frozen solution, respectively.

The excellent agreement between the results and Eq. (11) also rules out a significant contribution by a bimo-lecular singlet quenching process, for which $I_{DF}/I_{PF} \propto n_S n_T/n_S = n_T \propto (1+c_2 P)^{1/2} -1$. Finally, we note that the assignment of the DF in the PPVs to T-T annihilation is also in agreement with recent measurements of the frequency dependence of the triplet PA and ODMR [19].

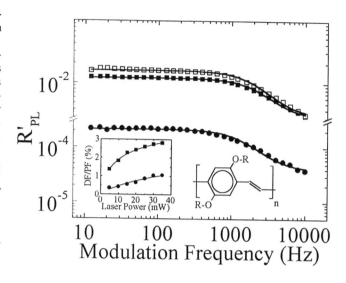

Figure 1. The DF R'_{PL} (see Eqs. (9) and (10)) of DHO-PPV powder (open squares), DOO-PPV film (closed squares) and < 0.025 mg/ml frozen DOO-PPV/toluene solution (circles) at 20 K vs the laser modulation frequency ν. The results are fit to a Lorentzian $A/[1 + \omega^2 \tau^2] + C$ yielding $\tau = 70$ μs in the powder and film and $\tau = 110$ μs in the frozen solution. Inset: The ratio of the DF to the prompt fluorescence (PF) vs laser power. The solid lines are the best fit to Eq. (11).

Figure 2 shows R'_{PL} vs ν in DBO-PPE powder and < 0.025 mg/ml frozen toluene solution at 20 K. While the behavior of the powder is in rough agreement with the simple Lorentzian with $\tau = 210$ μsec, the behavior of the frozen solution does not follow the best-fit Lorentzian obtained with $\tau = 520$ μsec. The poor agreement may be due to, among others, monomolecular triplet decay that is governed by two or more lifetimes τ_{Ti}. Indeed, very recent measurements indicate that irreversible processes occurring during photoexcitation of some PPV films at room temperature result in evolution of the DF from the behavior shown in Fig. 1 to behavior such as shown in Fig. 2.

The inset of Fig. 2 shows the similar dependence of I_{DF}/I_{PF} on P in DBO-PPE powder as seen in the inset of Fig. 1 for DOO-PPV film. It also shows that in this case I_{DF}/I_{PF} of the frozen solution is greater than that of the powder, in contrast to the weaker DF of the frozen DOO-PPV solution as compared to the film. This raises the possibility that some triplet formation may be related to specific sites or defects, consistent with their formation by back-transfer of triplet interchain polaron pairs [6] and the effects of room-temperature photodegradation of some PPVs mentioned above.

Figure 3 compares the DF and PF spectra of powder and solutions. It is interesting to note that while the DF spectra of the solutions are identical or only slightly red-shifted from the PF, they are clearly red-shifted in the powders and film. This $\Delta\lambda \sim 10$ nm red-shift, which corresponds to an energy shift $\Delta E \sim 30$ meV at $\lambda \sim 600$ nm, is consistent with the stronger dependence of the $1^1 B_u$ energy $E(1^1 B_u)$ on the conjugation length than that of the triplet $E(1^3 B_u)$:

This greater dependence of $E(1^1B_u)$ is due to its relatively extended size, as compared to the localized triplet [7,8, 20, 21]. Hence, in the longer π-conjugated segments of the solution $E(1^1B_u) < 2E(1^3B_u)$, but in the shortest segments resulting from strains and defects in the powder $E(1^1B_u) > 2E(1^3B_u)$.

Besides providing strong evidence for the presence and nature of the DF in π-conjugated polymers, the present results also impact our assessment of other

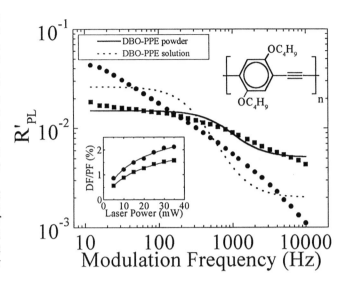

Figure 2. The delayed fluorescence (DF) R'_{PL} of DBO-PPE powder (squares) and <0.025mg/ml frozen DBO-PPE/toluene solution (circles) at 20 K vs the laser modulation frequency v (see Eqs. (9) and (10)). The lines are the least-squares Lorentzian $A/[1 + \omega^2\tau^2] + C$ fits with $\tau = 210$ μs for the powder and $\tau = 520$ μs for the frozen solution. Inset: The ratio of the cw DF to the prompt fluorescence (PF) vs laser power. The solid lines are the behavior predicted by Eq. (11).

processes in these materials. Thus, some ODMR and other studies [5,22,23] have suggested that nonradiative quenching of singlet excitons by polarons may be a significant channel competing with the PF. Yet the present results suggest that at the excitation intensities of this work bimolecular singlet quenching processes in the PPVs studied in this work, where

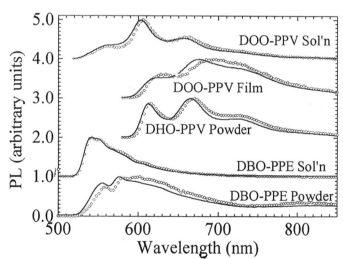

Figure 3. The uncorrected prompt fluorescence (PF) (solid lines) and delayed fluorescence (DF) (open circles) spectra of the PPV and PPE samples. Note the red shift of the DF relative to the PF in the film and powders.

agreement with the triplet-triplet annihilation model is excellent, are relatively inefficient.

SUMMARY

In summary, we have described frequency-resolved PL measurements of PPV and PPE derivatives which provide strong evidence for delayed fluorescence due to a bimolecular process, ostensibly triplet-triplet exciton annihilation to singlets. At 20 K the lifetimes of this process are 70 μsec in PPV powders, 110 μsec in their dilute toluene solutions, ~200 μsec in PPE powders, and ~500 μsec in their dilute toluene soutions. In PPV derivative films it contributes ~3% of the total emission at an excitation power of 35 mW.

ACKNOWLEDGEMENTS

Ames Laboratory is operated by ISU for the USDOE under Contract W-7405-Eng-82. This work was supported by the Director for Energy Research, Office of Basic Energy Sciences.

REFERENCES

1. *Proceedings of the International Conference of the Science and Technology of Synthetic Metals,* edited by Z. V. Vardeny and A. J. Epstein, Syn. Met. **84 – 86** (1997).
2. T. Azumi and S. P. McGlynn, J. Chem. Phys. **39**, 1186 (1963); R. G. Kepler, J. C. Caris, P. Avakian, and E. Abramson, Phys. Rev. Lett. **10**, 400 (1963); C. E. Swenberg, J. Chem. Phys. **51**, 1753 (1969).
3. E. M. Conwell and H. A. Mizes, Phys. Rev. B **51**, 6953 (1995); H. A. Mizes and E. M. Conwell, Syn. Met. **68**, 145 (1995).
4. N. F. Colaneri et al., Phys. Rev. B **42**, 11670 (1990); J. Rühe, N. F. Colaneri, D. D. C. Bradley, R. H. Friend, and G. Wegner, J. Phys.: Condens. Matter **2**, 5465 (1990).
5. W. Graupner, J. Partee, J. Shinar, G. Leising, U. Scherf, Phys. Rev. Lett. **77**, 2033 (1996).
6. V. Dyakonov, G. Rösler, M. Schwoerer, and E. L. Frankevich, Phys. Rev. B **56**, 3852 (1997).
7. L. S. Swanson, J. Shinar, and K. Yoshino, Phys. Rev. Lett. **65**, 1140 (1990).
8. L. S. Swanson, P. A. Lane, J. Shinar, and F. Wudl, Phys. Rev. B **44**, 10617 (1991).
9. X. Wei, B. C. Hess, Z. V. Vardeny, and F. Wudl, Phys. Rev. Lett. **68**, 666 (1992).
10. Q.-X. Ni et al., Syn. Met. **49-50**, 447 (1992).
11. L. S. Swanson, J. Shinar, Y. W. Ding, and T. J. Barton, Syn. Met. **55-57**, 1 (1993).
12. K. Pichler, D. A. Halliday, D. D. C. Bradley, P. L. Burn, R. H. Friend, and A. B. Holmes, J. Phys.: Cond. Matt. **5**, 7155 (1993); S. V. Frolov, M. Liess, P. A. Lane, W. Gellerman, Z. V. Vardeny, M. Ozaki, and K. Yoshino, Phys. Rev. Lett. **78**, 4285 (1997).
13. Y. Ding, Ph.D. Thesis, Iowa State University, Ames, IA, 1994 (unpublished).
14. J. Shinar, L. S. Swanson, Y. W. Ding, T. J. Barton, and Z. V. Vardeny, US Patent 5,352,906 (1994); J. Shinar, L. S. Swanson, F. Lu, and Y. W. Ding, US Patent 5,334,539 (1994).
15. C. Weder and M. S. Wrighton, Macromol. **29**, 5157 (1996).
16. The sample was the same as that used for the studies published in ref. 18.
17. Synthesized by M. De Long, Physics Dept., University of Utah, Salt Lake City, UT 84112.
18. E. L. Frankevich, unpublished results.
19. P. A. Lane, private communication.
20. D. Beljonne et al., J. Chem. Phys. **105**, 3868 (1996).
21. J. M. Warman et al., in *Optical Probes of Conjugated Polymers,* edited by Z. Valy Vardeny and L. J. Rothberg, SPIE Conf. Proc. **3145**, 142 (1997).
22. D. D. C. Bradley and R. H. Friend, J. Phys. Condensed Matter **1**, 3671 (1989); K. E. Ziemelis, A. T. Hussain, D. D. C. Bradley, R. H. Friend, J. Rühe, and G. Wegner, Phys. Rev. Lett. **66**, 2231 (1991).
23. Z. V. Vardeny and X. Wei, Mol. Cryst. Liq. Cryst. **256**, 465 (1994).

AUTHOR INDEX

SUBJECT INDEX